Target Earth

Target Earth

Meteorites, Asteroids, Comets, and Other
Cosmic Intruders That Threaten Our Planet

Govert Schilling

translated by Marilyn Hedges

The MIT Press
Cambridge, Massachusetts | London, England

The MIT Press
Massachusetts Institute of Technology
77 Massachusetts Avenue, Cambridge, MA 02139
mitpress.mit.edu

The MIT Press would like to thank the anonymous peer reviewers who provided comments on drafts of this book. The generous work of academic experts is essential for establishing the authority and quality of our publications. We acknowledge with gratitude the contributions of these otherwise uncredited readers.

This book was set in ITC Stone Serif Std and ITC Stone Sans Std by New Best-set Typesetters Ltd. Printed and bound in the United States of America.

Library of Congress Cataloging-in-Publication Data

Names: Schilling, Govert, author. | Hedges, Marilyn, translator.
Title: Target Earth : meteorites, asteroids, comets, and other cosmic intruders that threaten our planet / Govert Schilling ; translated by Marilyn Hedges.
Other titles: Doelwit aarde. English
Description: Cambridge, Massachusetts : The MIT Press, [2025] | Originally published in Dutch in 2022 under the title: Doelwit aarde | Includes bibliographical references and index.
Identifiers: LCCN 2024020069 (print) | LCCN 2024020070 (ebook) | ISBN 9780262551342 (hardcover) | ISBN 9780262381703 (epub) | ISBN 9780262381710 (pdf)
Subjects: LCSH: Asteroids—Collisions with Earth. | Impact craters. | Near-earth asteroids.
Classification: LCC QB651 .S3413 2025 (print) | LCC QB651 (ebook) | DDC 523.5—dc23/eng/20240929
LC record available at https://lccn.loc.gov/2024020069
LC ebook record available at https://lccn.loc.gov/2024020070

10 9 8 7 6 5 4 3 2 1

EU product safety and compliance information contact is: mitp-eu-gpsr@mit.edu

publication supported by a grant from
The Community Foundation for Greater New Haven
as part of the *Urban Haven Project*

Contents

Introduction

When I was a teenager and had a Saturday job in the local supermarket, I saved every cent I earned to buy my own telescope. After that I could be found every clear evening in the schoolyard next to my parents' house, looking up at the heavens. But whereas other amateur astronomers were mainly interested in binary stars, nebulae, and distant galaxies, I was fascinated by the Moon. It was a totally different world, its surface dotted with ink-black shadows from thousands of craters, large and small. I never tired of looking at it, especially once I understood that all those craters had been caused by cosmic impacts over the past few billion years. The surface of the Moon is one enormous disaster area.

Many of the Moon's craters measure tens or even hundreds of miles across. Small craters less than a mile in diameter were too minute for me to see with my amateur telescope. But even such a small scar in the crust of a celestial body is a very impressive sight. I experienced that awe in person for the first time in the summer of 2001, when I stood on the rim of Meteor Crater in Arizona: 175 yards deep, 50,000 years old, and created by the impact of a nickel-iron meteorite roughly 50 yards across.

Since then, I have visited a number of other meteorite craters in Europe, Australia, and South Africa. Every time, I try to imagine what it would be like to witness such a catastrophic cosmic collision—preferably at a safe distance, of course. I've never managed to do that; the universe is beyond our powers of imagination even on this relatively modest scale. But while serious impacts are thankfully rare, they have nonetheless affected the evolution of life on Earth in the past and will continue to cause death and destruction in the future.

In recent years, astronomers have been getting increasingly good at tracking potentially dangerous cosmic projectiles. Recently NASA even succeeded in deliberately deflecting a small asteroid. This is a far cry from the currently (and perhaps always) inconceivable idea of preventing a possible imminent impact from a 6-mile-wide celestial body like the one that wiped out the dinosaurs 66 million years ago. But in the future, we may manage to avert the destruction of a city or a small country if such an extraterrestrial natural disaster should ever be at risk of occurring.

This book will tell you about more or less all aspects of cosmic impacts. From small meteorites to devastating collisions; from the extinction of the dinosaurs to impact craters on other celestial bodies; and from searches for near-miss lumps of rock to ways of protecting humanity from an assault from the cosmos. And along the way you will read about the differences between asteroids and comets, about near misses in the past and in the future, and about the positive flipside of all these visitations from space: if our planet had not been the target of cosmic rubble since its very formation, life on Earth would probably never have begun.

As humans, we know only too well that we are, to some extent, at the mercy of nature's whims. Earthquakes, volcanic eruptions, and tsunamis take their toll with due regularity. But not everyone realizes that cosmic impacts also belong on that list of terrifying natural disasters. This book will bring you right up to date—not to give you sleepless nights, but to make you realize that we cannot separate our own existence from the universe around us.

1
Falling Debris

I first saw Ann Hodges when I was about 8 years old. It was in a black-and-white photograph in the beautifully illustrated book *The Earth*, part of the Time Life series that my parents had on their bookshelf.[1] Hodges lay on her back in bed with her eyes closed, stone dead or so it seemed. A doctor wearing glasses and a bow tie lifted her nightdress a little, revealing a large, dark wound on her left hip. The caption underneath described her as the victim of a 10-pound meteorite. As a young boy, I found it horrifying; the photo will be forever etched in my memory.

Hodges, age 34, from Sylacauga, Alabama, was not actually dead. But on November 30, 1954, around midday, a rock from space penetrated the atmosphere and shot through the roof of her wooden house at 125 miles per hour, hitting her while she napped on the sofa. After some time in the possession of the owner of the rental house, the meteorite is now in the Alabama Museum of Natural History in Tuscaloosa.

Rocks that fall to Earth from space: A couple hundred years ago, anyone who believed such a phenomenon would have been declared insane. That all changed in 1803. French astronomer Jean-Baptiste Biot traveled from Paris to the town of L'Aigle in Normandy, where a veritable shower of rocks apparently rained down on April 26 of that year. Numerous eye-witness reports and a study of some 2,000 fragments convinced Biot that this was no hoax: the cosmos really was raining down on Earth.[2]

Ernst Chladni—a German physicist and musician known for his research on acoustic vibrations—had already suggested this in 1794, after Barbotan in southern France was also hit by falling debris on July 24, 1790. And what about those reports, recorded in the official history of the Ming dynasty, of a catastrophic natural disaster in the Chinese district of Ch'ing-yang in the spring of 1490? "Rocks fell like rain," according to the chroniclers. Some sources report many thousands of victims.[3]

Given this history, Theresa Davies and Kathleen Clifton had a lucky escape. On September 30, 1984, the two Australian friends were sunbathing on Binningup Beach in Western Australia when their relaxed afternoon was shattered by a loud whooshing sound, a dull thud, and a spray of sand. A 1-pound meteorite had landed right beside them. On October 9, 1992, Michelle Knapp from Peekskill, New York, the proud 17-year-old owner of a 1980 Chevrolet Malibu, heard a loud bang: her prized car, which had been parked just in front of the door to her parents' house, had been hit by a space rock weighing 28 pounds. A month later, on November 8, a house in Wethersfield, Connecticut, was struck by a meteorite. This was particularly remarkable because something similar had also happened in 1971 in the same town, just over a mile away.

If Earth is so clearly in the cosmic firing line, it is actually a miracle that there have been so few casualties. That report of thousands of deaths in China has never been officially confirmed, any more than the legend of an Italian monk allegedly fatally struck by a meteorite in 1511, or the rumor about an Indian bus driver who suffered the same fate in 2016. There are even doubts about the authenticity of the well-known story of a street dog in Egypt that was killed on June 28, 1911, after being hit by a space rock. In fact, there is only one truly reliable account of "death by meteorite": an Iraqi man who did not survive a shower of rocks near Sulaymaniyah on August 10, 1888.

Despite being small, my home country of the Netherlands has also not been spared. On April 7, 1990, a meteorite weighing just under 2 pounds smashed through the tiled roof of a house in Glanerbrug, near Enschede. The residents found numerous pieces of debris in the attic, the largest of which weighed 4.8 ounces. The house was demolished in 2005, and today the site of the meteorite impact is marked by a simple plaque. And on January 11, 2017, a rock weighing some 1.1 pounds crashed through the roof of a garden shed in Broek in Waterland, just north of Amsterdam.

The Glanerbrug and the Broek in Waterland (meteorites are named after the place where they come down) are the most recent Dutch specimens. The smallest Dutch meteorite (2.4 ounces) is the Diepenveen, the existence of which only came to light by chance this century. Farm laborer Albert Bos and his wife saw the rock fall on October 27, 1873, and took it to the local headmaster. By a roundabout route, amateur astronomer Henk Nieuwenhuis came across the rock in 2012. He realized straightaway that it was a real meteorite.

So meteorites are not so very rare. Around the world, tens of thousands have now been found, with a combined mass of several hundred tons. You come across them in geological or natural history museums, at exhibitions in planetariums or public observatories, and—in large numbers—on the internet. On websites like eBay and Catawiki there is a lively trade in large and small space rocks. Some people wear a meteorite around their neck as jewelry; I have a small iron meteorite keychain. Of course, there are also plenty of scams; even false certificates of authenticity are traded, so you do need to be a bit careful.

Most meteorites found in museum collections are iron meteorites, composed mainly of iron and nickel—the most common metals in the cosmos. My keychain specimen is also an iron meteorite. Yet "siderites," as they are also called, make up less than 6 percent of all space rocks pelting Earth; stony ones are far more numerous. Because of its higher porosity, however, a stony meteorite is much less likely to survive the journey through Earth's atmosphere, and, once it hits the ground, it naturally weathers much faster.

How can you recognize a meteorite? With a stony meteorite, the melting crust—a thin, black layer formed due to friction with the atmosphere—is an important clue. The vast majority of stony meteorites also contain chondrules—spherical inclusions no larger than a tenth of an inch—thus leading to these meteorites also being known as chondrites. Iron meteorites can be recognized by their magnetism, usually, and small "indentations"— regmaglypts ("thumbprints")—that form during the descent through the atmosphere. If you cut an iron meteorite in half and place it under a microscope, you will often see long, perfectly straight iron-nickel crystals, known as Widmanstätten patterns.

The chances of finding a random meteorite yourself are incredibly small. Most of the "weird rocks" that are found (I get a notification in the mail almost once a month) turn out on closer inspection to be ordinary terrestrial specimens, or what is known as iron slag, which comes from the metal industry and is used to pave roads. The website https://geology.com/meteorites/meteorite -identification.shtml can help you decide whether you have found something unusual. If you think you have, you can contact professional geologists, but you should be prepared for a disappointing response.

Obviously, the easiest way to find a meteorite is to actually see it come down, or to locate it after it causes damage to a house or car, as in the cases mentioned previously. But a bright trail of light in the sky can also be a clue. If it is photographed from different directions, its three-dimensional trajectory through the atmosphere can be reconstructed and you can calculate where any debris may have landed. For example, over 10 ounces of meteorite material was recently recovered near Winchcombe in the UK: astronomers analyzed photos of a bright fireball visible on February 28, 2021, and appealed to people living near the so-called strewn field to see if they could find anything.

Dutch American astronomer Peter Jenniskens (from the SETI Institute in California) has some experience in this. In late 2008, he set out with scientists and students from the University of Khartoum to search for "fresh" meteorites in the sweltering Nubian Desert after a telescope in Arizona had detected a cosmic rock a few yards in size that was due to "collide" with Earth on October 7 of that year (just one day after the discovery). Based on the images, they could calculate the exact area where any meteorites would have come down. Indeed, during three expeditions, as many as 280 fragments were found!

Ten years later, Jenniskens had another success. On June 2, 2018, the same Arizona telescope observed a small projectile racing toward Earth, and that same evening the heavens above Botswana were lit up by an exceptionally bright fireball. Three weeks later, Jenniskens found the first meteorite in the extensive Kalahari Game Reserve (named Motopi Pan, after a nearby watering hole) while armed park rangers stood guard against some overly curious lions. In the following months, more than 20 fragments were discovered.

NASA astronaut Stan Love is another person who knows just how exciting it is to find an extraterrestrial rock—not on board the International Space Station that was his home for two weeks but in Antarctica, where I met him in December 2012. Love was a member of an ANSMET (Antarctic Search for Meteorites) expedition during which he and his fellow team members rode on ice scooters in formation across the frozen continent looking for conspicuous dark rocks on the blindingly white surface of the ice. Meteorites that have fallen to Earth over tens of thousands of years are forced upward and concentrated at the foot of the Transantarctic Mountains by the glacial movement of Antarctic ice. Success assured.

Anyone who still wants to go meteorite hunting needs only look in their own gutters, as Norwegian guitarist and artist Jon Larsen did in 2016. Larsen became intrigued by the idea that micrometeorites are also constantly raining down on our planet—totaling over 10 tons every 24 hours, scattered all over Earth's surface. He collected dust particles from his gutter, used a magnet to select possible extraterrestrial candidates, and was indeed able to identify a number of micrometeorites under the microscope: small, dark spheres with light spots. Amateur geologists and astronomers have also found micrometeorites in the same way.

It should be clear by now that extraterrestrial matter is constantly raining down on Earth in the form of microscopic dust particles, space debris, tiny fragments of rock and iron, fist-sized meteorites, whoppers weighing many pounds, and sometimes even complete boulders. And, as always in nature, the smaller examples—the micrometeorites and harmless dust particles—are far more numerous than the larger ones, such as the 28-pound projectile that hammered into the trunk of Knapp's Chevy Malibu. That's just as well because if a really heavy object falls on Earth, the consequences will be disastrous.

In November 2016 I went to see the world's biggest meteorite. Weighing over 60 tons, it is so heavy that it has never been transported to a museum. Instead, it remains where it landed more than 80,000 years ago on the Hoba farm near Grootfontein in northern Namibia. The Hoba meteorite is an enormous lump of iron and nickel a couple of yards in diameter. It is an impressive monster, and definitely not something you want to have land on your roof.

But that's not something we can control.

2
Names and Numbers

In the previous chapter you got to know about stony meteorites (around 93 percent of all space rocks) and iron meteorites (6 percent). And that remaining 1 percent? These are the rare stony iron meteorites that contain both rocks and metals. Stony meteorites, in turn, are divided into chondrites (with small, spherical inclusions) and achondrites (without those distinctive chondrules), which are about 10 times as rare. The latter group also includes the unusual HED meteorites (howardites, eucrites, and diogenites), to which the Motopi Pan from Botswana belongs. The mineralogical composition of the HED meteorites shows that they originate from the large asteroid Vesta.

You must have noticed that astronomers and geologists like there to be a box for everything, and everything to be in its box: they have separate names for every object. An astronomer will correct you in no uncertain terms if you say you saw a meteorite appear in the night sky. The term meteorite is only used for natural objects from space that have landed on Earth. The light phenomenon that you might see in the night sky is officially known as a meteor. It is often referred to as a "shooting star," but that too is a description that is open to criticism from professional quarters, because stars of course don't really fall to Earth.

You often read that meteors result from space rocks and dust particles that burn up in Earth's atmosphere. In reality, things are more complicated than that: the dust or grit particle (usually no more than a tenth of an inch in size) penetrates the upper, tenuous layers of the atmosphere at a speed of at least 6 miles per second. Due to the heat created by friction with the oxygen and nitrogen molecules in the atmosphere, the rock will, in most cases, completely evaporate—not burn up!—while the air molecules will enter an excited state. When they revert to their so-called ground state shortly afterward, they emit light. This is how a meteor is created, at an altitude of 50 to 55 miles above Earth's surface.

It is possible to see several meteors every clear, moonless night: these are short streaks of light in the night sky that often last no more than a second. But apart from these sporadic meteors, there is occasionally a meteor swarm or meteor shower, and dozens of shooting stars can be seen every hour. This happens when Earth is moving through a vast cloud of cosmic debris and dust on its annual journey around the Sun. Seen from the ground, the resulting meteors appear to come from a single region in the sky; the shower is named after the constellation in which that so-called radiant is located.

The best-known meteor shower is the Perseid shower, named after the constellation Perseus, that makes its appearance every year and peaks around August 12–13.

As has been said, most meteors are caused by tiny particles, similar to the hail of pellets from a shotgun. Such a tiny chunk of matter naturally evaporates completely during its journey through the atmosphere. This produces a relatively faint meteor, and no meteorite lands on Earth. Even a projectile the size of a hefty pebble will not manage to reach Earth's surface, although the accompanying meteor will of course be much brighter, and visible over a longer trajectory. Only if the incoming object is the size of a grapefruit is there a chance of a meteorite falling to the ground. The dazzling light phenomenon associated with such a descent is called a fireball or bolide.

Besides the familiar terms "meteor" and "meteorite," there is a third related term in the astronomical dictionary: meteoroid. This is the name given to a space rock before it lands on Earth. To recap: meteors (and fireballs) are caused by a meteoroid entering Earth's atmosphere, and, if it does not vaporize completely during its journey through the atmosphere, a meteorite is the rock that lands on Earth's surface.

Even astronomers themselves sometimes get confused with all these terms and designations, as was the case with the meteorites found in 2008 and 2018 by Peter Jenniskens and his colleagues. They were fragments of small celestial bodies that had been discovered shortly before impact by the Catalina Sky Survey—a search program at an observatory in Arizona. So, were these celestial bodies extremely large meteoroids a few yards in size, or small asteroids? In any case, since they were discovered when they were still in interplanetary space, they did receive an official asteroid designation: 2008 TC$_3$ and 2018 LA, respectively.

Asteroids are celestial bodies, enormous numbers of which orbit the Sun like miniature planets, but they are much too small to warrant the term planet. To give you an idea: the smallest planet in the solar system, Mercury, is about 3,000 miles in diameter; most asteroids are no more than a few tens of miles across. (The biggest, Ceres, is 588 miles in diameter; there are in fact only around 200 known asteroids larger than 60 miles.)

Ceres was also the first to be discovered, by Italian astronomer Giuseppe Piazzi, on January 1, 1801. It was quickly followed by the discovery of Pallas, Juno, and Vesta, by astronomers in Germany. Piazzi and his German colleagues, who had been excited by the discovery of the distant planet Uranus in 1781, welcomed the four newcomers as new planets. But when astronomers found many more of these small objects in the mid-nineteenth century, they were put into a separate class—partly on the initiative of the discoverer of Uranus, William Herschel, who coined the word "asteroid."

At the present time there are over a million known asteroids. Most of them are just a few miles in size, and they have been discovered only in recent years by big automatic search programs using powerful telescopes, like the Lincoln Near-Earth Asteroid Research project (LINEAR) in New Mexico and the Catalina Sky Survey in Arizona mentioned before. Over a million sounds very impressive, but all these asteroids together represent only 3 percent of the mass of the Moon! At one time, people thought that asteroids might be fragments of an exploded planet, but astronomers now know that they are rocky remnants from when the solar system was formed.

Once an asteroid has been discovered, it is given a temporary name made up of the year when it was discovered, followed by two letters (and sometimes another number in subscript). When its precise orbit through the solar system has been determined, the heavenly body is given a definitive number (officially always written between brackets); the catalogue of the Minor Planet Center of the International Astronomical Union already includes more than 600,000 numbered asteroids. Many of them (some 23,500 as of October 2022) also have an official name.

By far most asteroids are in a wide belt between the orbits of the planets Mars and Jupiter, on average a few hundred million miles from the Sun, and they have orbital periods lasting several years. But many have a divergent orbit, regularly moving through the inner regions of the solar system, and they can come dangerously close to Mars or Earth. Depending on their precise orbital characteristics, these are referred to as Apollo,

Amor, Aten, or Atiras asteroids. In addition, almost 10,000 asteroids have been discovered moving in more or less the same orbit around the Sun as the giant planet Jupiter: these are known as Trojans.

Given that more than a million asteroids have been discovered, it comes as no surprise that some have very exceptional characteristics, or very unusual orbits. Take, for instance, the nameless object that revolves on its axis extremely slowly: once every 78 days. A very different specimen is 2014 RC (just 13 yards in size), which has a rotation period of no more than 16 seconds! The object 2021 PH_{27}, discovered in August 2021, has the shortest known orbital period around the Sun: 113 days. And 2010 EQ_{169} moves in an orbit that is more or less perpendicular to the central plane of the solar system while 2013 LA_2 revolves in the "wrong" direction around the Sun, like a cosmic ghost driver. In short: it is a motley crew.

And asteroids are not the only small inhabitants of the solar system. In addition to these rocky celestial bodies, there are also countless icy comets orbiting the Sun, mostly in extremely elongated trajectories. Again, these objects are a few miles in size, consisting mainly of ice: frozen water, but also frozen gases like carbon monoxide, carbon dioxide, methane, and ammonia. If a comet traverses the inner regions of the solar system, some of that ice evaporates, releasing dust and grit particles. The small nucleus of the comet envelops itself in a huge cloud of gas and dust particles (known

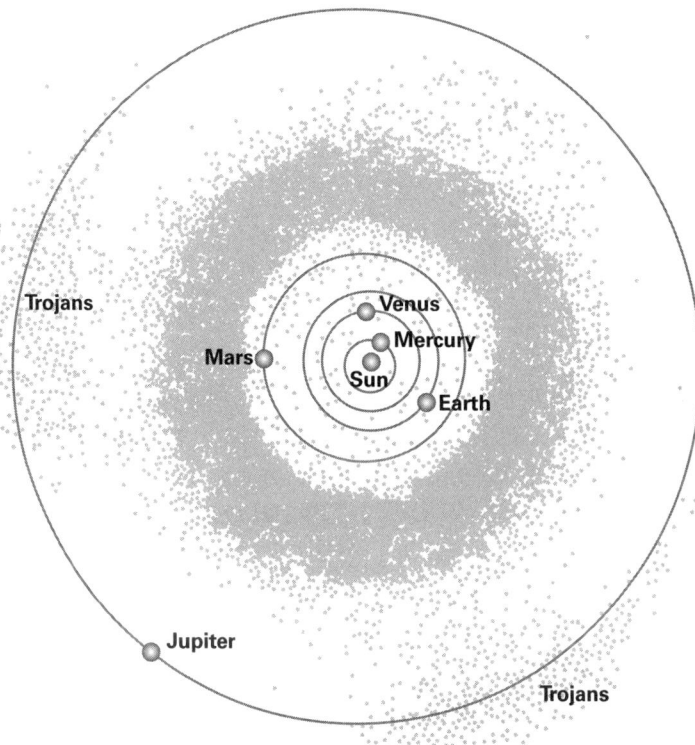

Figure 2.1

The distribution of the largest asteroids in our solar system. By far the most are located between the orbits of Mars and Jupiter. There are also many "Earth-grazers," as well as two groups of asteroids moving around the Sun in the same orbit as Jupiter. Source: Wikimedia Commons.

as the coma), and when that material is blown away by the solar wind (electrically charged particles blown into space by the Sun), a comet exhibits a long and distinctive tail.

A comet's tail is extremely tenuous but can easily reach a length of up to tens of millions of miles. The gas atoms in the tail become ionized through interaction with the solar wind, and as a result emit radiation; the dust particles reflect the sunlight. All in all, this makes for an impressive sight in the night sky, with the comet's tail always pointing away from the Sun, of course. (That means, by the way, that a comet also sometimes moves "against its tail," namely when it recedes from the Sun again.)

Comets have been observed since time immemorial. They were generally thought to be a bad omen. Greek astronomers even assumed that they were phenomena in Earth's atmosphere. In 1577, the Danish astronomer Tycho Brahe was the first to show that comets are much farther away than previously thought; Edmond Halley discovered in the early eighteenth century that some orbit the Sun in elongated elliptical orbits. Comet Halley, for example, which is named after him, has an orbital period of 76 years, and will be visible in the sky again in 2061.

Bright comets that you can see with the naked eye are relatively rare—the last one was comet Neowise (named after the space probe that discovered it), in the summer of 2020. About 5,000 comets are currently known, but their actual number is far greater. During the vast majority of its orbital period, a comet is at a very great distance from the Sun and is simply far too faint to be seen from Earth. On top of that, most comets have orbital periods of many thousands or even hundreds of thousands of years—Halley, with its 76 years, is one of the short-period comets.

Incidentally, there is a special relationship between comets and meteor showers. In 1866, Italian astronomer Giovanni Schiaparelli discovered that dust particles from the Perseid shower move through the solar system in the same orbit as comet Swift–Tuttle. The latter orbits the Sun in an elongated orbit once every 133 years. With each passage, gas and dust particles are released from the porous nucleus of the comet, and these get distributed throughout the orbit over time. Something similar applies to most other meteor showers: the Orionids (seen around October 22) and the Eta Aquariids (early May) are both associated with Halley's Comet.

With eight planets, a few hundred planetary moons, over a million asteroids, and thousands of comets, the inventory of our solar system is still not complete. Beyond the distant planet Neptune, numerous Kuiper belt objects still orbit the Sun—largely frozen celestial bodies that include dwarf planet Pluto. Some of these so-called ice dwarfs linger temporarily in the region between the orbits of Saturn and Neptune; astronomers then refer to them as the Centaurs. At a gigantic distance of trillions of miles, the Sun is also surrounded by the Oort cloud, which contains trillions of small cometary nuclei, while interstellar objects, such as the remarkable celestial body 'Oumuamua, discovered in October 2017, are constantly flying through the solar system at high speed.

Incidentally, nature pays little heed to the human love of giving things labels: for some objects, it is hard to tell whether they are asteroids or comets. However, all these small celestial bodies have one thing in common: if their orbit crosses that of Earth, they could in principle collide with our planet. That has happened in the past, and it will certainly happen again in the future. Earth is a cosmic target, and every now and then there is a collision. The numerous scars on our planet's surface bear witness to that.

3
Cosmic Scars

In summer 2021, Chinese scientists published an article in the journal *Meteoritics & Planetary Science* in which they announced the discovery of an impact crater that had never previously been recognized.[1] It is located near the town of Yilan, in Heilongjiang Province, in northeastern China. The crater is 1.1 miles in diameter and a few hundred yards deep and is estimated to have been created between 50,000 and 100,000 years ago, when a cosmic projectile many tens of yards across collided with our planet.

It is not surprising that the Yilan crater had never previously been identified. It is highly eroded and is located in a densely forested area. It is a very different situation from the slightly smaller Meteor Crater in Arizona (0.75 miles in diameter). This crater is younger—about 50,000 years old—and is in a bone-dry desert environment where there is far less erosion. Meteor Crater is the best-preserved impact crater on Earth. If you ever happen to be in the area—about 35 miles east of Flagstaff—be sure to take a look.

Geologist and businessman Daniel Barringer was convinced early in the twentieth century that Meteor Crater (also known as Barringer Crater) must have been caused by a cosmic impact, while most scientists at the time still assumed its origin was volcanic, mainly because Flagstaff is located in an area rich in volcanic formations. It was not until 1960 that geologist Eugene Shoemaker discovered shocked quartz crystals in the rock, which can only form during a sudden, violent explosion. No one now doubts that Meteor Crater is the scar left when a roughly 50-yard iron meteorite slammed into Earth at high speed.

North America was not inhabited by humans 50,000 years ago, but for the mammoths and giant sloths that populated the area at the time, the impact must have been catastrophic. At the point when a large meteorite slams into Earth's surface, all its kinetic energy is converted into heat. The result is a colossal explosion, like a nuclear bomb going off but without the radioactivity. This is also the reason why impact craters are so similar to bomb craters, and that in almost all cases they are perfectly round and symmetrical.

In the 1960s, Shoemaker, who worked for NASA at the time, made smart use of this similarity between impact craters and bomb craters to train Apollo astronauts Neil Armstrong and Buzz Aldrin. He used explosives to create an almost exact replica of the Apollo 11 landing area in the vast cinder fields at the base of Sunset Crater (a volcanic crater near Flagstaff). The many dozens of small craters near the planned landing site of the *Eagle* lunar lander, measuring between 2 and 15 yards across, were reproduced as accurately as possible.

Major impacts like those that created the Yilan crater and Meteor Crater are fortunately rare. But smaller meteorite impacts can also leave craters behind. On September 15, 2007, for example, near Carancas in southeastern Peru (close to Lake Titicaca), a crater 14 yards in diameter and 5 yards deep was created by the impact of a sizeable stony meteorite. The space rock was probably around 3 yards in size when it entered the atmosphere, but it disintegrated during its descent.

This also happened to the Chelyabinsk meteorite, which came down over the Russian city of the same name in the early morning of February 15, 2013, creating a spectacular light show there, captured by numerous dashcams. The built-up air pressure caused the 19-yard chunk of rock to explode at a height of several miles. The resulting shockwave shattered numerous windows in the city, and more than a thousand people suffered cuts from flying glass shards. The largest fragment of the meteorite was later fished out of a nearby frozen lake; it was a yard and a half in size and weighed 1,000 pounds. Had this piece of rock come down on land, it would undoubtedly have created a sizeable crater.

More than a century earlier, on June 30, 1908, Russia was also hit by a huge meteorite (or possibly a fragment of a comet) measuring between 50 and 60 yards in diameter. It flew through the atmosphere at enormous speed, and exploded at high altitude over the Tunguska River, in a virtually uninhabited area of Siberia. The blast wave from the atmospheric explosion razed 1,200 square miles of pine forest to the ground. According to some researchers, the approximately 450-yard-wide and almost circular Lake Cheko marks the crater that would have been created by the impact of a 10-yard fragment, but this has never been definitively confirmed.

A major cosmic impact is different from the descent of an average meteorite. Small meteorites are always slowed down as they pass through Earth's atmosphere, and thus ultimately fall to the ground at relatively low speed. A bigger rock—larger than a few yards in diameter—retains much of its original speed, which is at least 6 to 12 miles per second. The kinetic energy is great enough for a crater to be formed, and the direct and indirect effects are felt far and wide. Such a rare phenomenon must have made an enormous impression on our forebears.

The Henbury craters in Australia—a group of 14 craters measuring between 7 and 200 yards, approximately 75 miles west of Alice Springs—were formed around 2700 BCE, when the region was inhabited by Indigenous Australians. The craters are still regarded as holy ground for the local Arrernte people. The same is true for Kaali, a group of nine craters on the island of Saaremaa (off the coast of Estonia), the largest of which is 120 yards in diameter and contains a still-water lake. The Kaali crater field dates from around 1500 BCE and must have been formed by the impact of the fragments of a cosmic projectile weighing tens of tons.

What are the chances that such a violent impact could take place again today or tomorrow? Fortunately, it is highly unlikely. A Chelyabinsk-like stony meteorite of around 20 yards across only lands on Earth once every 50 years or so. An impact like Tunguska (55 yards in diameter) occurs once every 750 years or so. These kinds of objects are not large enough to reach the surface of Earth intact: the projectile explodes into numerous fragments in the atmosphere. (Except, of course, when it is an iron meteorite, as in the case of Meteor Crater, but these are more than 15 times as rare.)

Heavier impacts that can cause a crater are far less frequent. Roughly once every 5,000 years Earth is hit by a rocky celestial body 100 yards across, and a crater is formed measuring half a mile in diameter. Projectiles of 450 yards in diameter collide with our planet roughly once every 100,000 years and leave behind a scar of some 3 miles in size. And a half-mile-sized lump of rock strikes just once every 500,000 years, forming a crater more than 5 miles in diameter.

The consequences of such cosmic impacts can all too easily prove disastrous—not only in the area where the crater forms but for miles around due to the energy from the explosion and the resulting shockwave. A projectile 450 yards in diameter can reduce a state the size of Texas to ashes, and one measuring half a mile in diameter would annihilate an entire continent. Incidentally, for

Figure 3.1
Meteorite craters identified on the surface of Earth. Most have been
found in places where the surface consists of ancient terrestrial crust.
Source: impact-structures.com.

the largest objects it does not matter much whether the
impact occurs above land or sea—tsunamis of a hundred yards in height are not something anyone wants.

Homo sapiens has only been on Earth for a few hundred thousand years, and the written history of mankind
barely covers some 5,000 years. It would seem that—so
far—we have managed to avoid such disasters, although
there are people who believe that particular myths and
folklore (such as the Biblical story of the destruction of
Sodom and Gomorrah) can be traced back to cosmic

impacts. But in the distant geological past—millions or even billions of years ago—catastrophic impacts occurred with some regularity, leaving colossal scars on Earth's surface.

One example is the Nördlinger Ries in southern Germany, a gigantic crater measuring 15 miles across. From the church tower of the picturesque little town of Nördlingen, almost in the center, you can see the rim of the crater all around you, at least if you know what to look for. The impact took place around 15 million years ago and was caused by an asteroid measuring around 1 mile in diameter. It was probably a "double asteroid" because the nearby Steinheim crater, a mere 2.4 miles in diameter, dates from the same time.

Almost as large—but more difficult to recognize in the landscape—is the 14-mile-wide crater near Rochechouart in the French Massif Central. It is estimated to be 200 million years old and, like Meteor Crater and Nördlinger Ries, it was not recognized as an impact structure until the 1960s. And in Sweden there is the Siljan Ring, Europe's largest impact crater. It measures 32 miles in diameter and is around 375 million years old.

The Siljan Ring is easy to find on satellite photos (or on Google Maps) because the rim of the crater is marked by a number of lakes. Intuitively, you might expect one big lake right in the center, but large impact craters often have a central elevation. This so-called uplift comes about because partly melted rock from the newly formed high rim of the crater flows inward again and is pushed upward at the center. The 214-million-year-old Manicouagan crater in Quebec, Canada, is also outlined by a ring-shaped lake some 45 miles in diameter.

In most cases, old impact craters have been strongly eroded by wind, water, and vegetation, or they are even completely "erased" by geological processes that are constantly at work on the surface of our planet. It is only in the most ancient parts of Earth's crust that the scars of cosmic collisions can still sometimes be identified. Take, for example, the Sudbury crater in Ontario, Canada. When this was formed 1.85 billion years ago, it had an estimated diameter of 80 miles. Now, all that is visible is a strongly misshapen depression in the landscape where a lot of iron and nickel are mined, both of which can, in a certain sense, be termed extraterrestrial metals.

An even bigger and older structure is the Vredefort crater in South Africa. Two billion years ago, a celestial body measuring 6 to 10 miles in diameter crash-landed there, resulting in a crater that must have been between 100 and 200 miles in size. All that is visible today are the concentric mountain structures near the center, known as the Vredefort Dome, intersected by the Vaal River. In 2016, I camped on the banks of the river, in the middle of the world's largest impact crater with a view of hilltops that, after 2 billion years, still bear witness to this devastating catastrophe.

Two billion years ago, life on Earth was just in its infancy: there were simple single-celled organisms swimming and swarming around, and we have no idea how and to what extent they suffered from the force of the impact. But the last time Earth was affected by a cosmic object around 6 miles in size, the situation was completely different. This impact—possibly the best known of all such events—left behind a 110-mile crater that is now hidden from view beneath thick layers of sedimentary rock. The Chicxulub crater is the tangible signature of the natural disaster that brought an end to the reign of the dinosaurs 66 million years ago.

4
Death of the Dinosaurs

It's not something you think about every day, but we owe our existence to a catastrophic cosmic impact. Some 66 million years ago—10 times farther back in time than the common ancestor of humans and chimpanzees—an unimaginable natural disaster put a permanent end to the dinosaurs. This suddenly opened new, rapid evolutionary possibilities for mammals, ultimately including primates. The rest is history, as the saying goes: if the dinosaurs hadn't become extinct, *Homo sapiens* would almost certainly never have walked the earth.

Dinosaurs appeal enormously to the imagination. It is a fascination that is much older than the blockbuster film *Jurassic Park*: even as a small boy in the mid-1960s, I read all the library books on dinosaurs I could find and knew their names by heart. The first fossilized remains of these giant reptiles were found at the start of the nineteenth century, even before Charles Darwin penned his theory of evolution in his 1859 book *On the Origin of Species*. We now know that dinosaurs first appeared on the scene some 230 million years ago, and from just under 200 million years ago were the dominant life form on Earth. They might still have been so today had the cosmos not put a stop to that.

We know from geological research that dinosaurs became extinct 66 million years ago, at the threshold of the Cretaceous and Paleogene periods. These time periods are often defined by the fossils found in ancient rock layers, so it is no coincidence that the disappearance of the horned *Triceratops* and the dreaded *Tyrannosaurus rex* coincided with such a geological transition. At the same time, countless other life-forms also disappeared, both on land and in the ocean. In total, three-quarters of all biological species on Earth perished. The term "mass extinction" is therefore no exaggeration.

In 1979, at a geological conference, Luis and Walter Alvarez (father and son) came up with a controversial theory: the mass extinction at the Cretaceous–Paleogene boundary (the K–Pg boundary) was likely caused by the impact of a 6- to 10-mile-wide asteroid. The two scientists had discovered an increased concentration of iridium, a metal that is thought to have an extraterrestrial origin, in a thin 66-million-year-old clay layer found in Italy. Together with Frank Asaro and Helen Michel, they published their theory on June 6, 1980, in the leading scientific journal *Science*.[1]

Iridium is a heavy and rare element. The heat from Earth's formation caused it to sink—along with iron and nickel—to the core of the molten planet; it is virtually absent in Earth's crust. Most asteroids never underwent such a process of differentiation, so iridium is much more homogeneously distributed in these small celestial bodies, making its concentration far higher than in Earth's crust. After a catastrophic impact (where the celestial projectile is completely vaporized and blasted into the atmosphere), a thin layer of asteroidal material, recognizable from this increased iridium abundance, will rain down almost everywhere on Earth's surface.

Incidentally, the honor of this discovery could have fallen to Dutch geologist Jan Smit of the Free University in Amsterdam. Smit had previously investigated the sudden extinction of foraminifera (microscopic marine organisms) at the Cretaceous–Paleogene boundary. In the mid-seventies, he discovered increased concentrations of other heavy metals in clay strata from Spain, although his analysis did not include a specific search for iridium. Smit's article in *Nature* appeared on May 22, 1980, two weeks before the article by his American colleagues (or competitors).[2] It did not take long before the impact theory reached the popular press and the general public.

It is impossible to put into words exactly what happened on that disastrous day 66 million years ago. A colossal piece of cosmic rock, between 6 and 10 miles in diameter and weighing an unbelievable couple of trillion tons, came hurtling toward Earth at a speed of 12 miles per second. The enormous mass and speed of the projectile meant it had a tremendous amount of kinetic energy: more than a billion times the explosive energy of an atomic bomb (for the number nerds: more than 1 yottajoule). When the asteroid came to a halt on—or rather, partially in—Earth's crust, all that energy was converted into heat.

The consequences were apocalyptic. Pulverized and vaporized asteroid material was hurled into the stratosphere, along with 10 times as much molten rock from Earth's crust. Glowing debris fell back to Earth, spreading far and wide and causing giant forest fires. In no time at all, the finer particles spread all over the planet. For at least 10 years, the Sun was obscured by large amounts of dust in the atmosphere. The average temperature on Earth plummeted at least 40 degrees Fahrenheit. Photosynthesis came to a virtual standstill, food chains were broken, and eventually almost all life on Earth gave up the ghost.

It sounds like a nightmare scenario, which is just what it was. The scientific community in the early eighties was still by no means convinced by the impact story. Did all those animal species really disappear so suddenly from the scene? Could the dinosaurs not have become extinct as a result of "natural" climate changes on Earth? According to many geologists and paleontologists, the volcano theory also had a lot going for it, given that the transition from Cretaceous to Paleogene is characterized by extremely high volcanic activity. Who knows, it could have been the dust and carbon dioxide emitted by volcanoes that eventually killed off the dinosaurs.

That fierce volcanic activity took place mainly on the Indian subcontinent. A small tectonic plate shifted north at a considerable speed of some 8 inches per year and collided with the large Eurasian plate (the "crumple zone" of that collision is today's Himalayan mountain range). Sixty-six million years ago, India was above the hot spot in Earth's mantle that now fuels the volcanic activity on the French island of Réunion. But at that time significantly more magma welled up from the planet's interior: within the space of a million years, India was buried by as much as 120,000 cubic miles of flood basalt—enough to cover the state of Connecticut with a 20-mile-thick layer. The creation of this volcanic feature, called the Deccan Traps, certainly also had a profound impact on the evolution of life on Earth.

However, some 10 years after the publication by the Alvarezes and their team, as well as that by Smit, nobody could ignore the asteroid impact theory any longer. A 110-mile crater of just the right age was discovered below the coast of what is now Mexico's Yucatán Peninsula. Actually, that subterranean circular structure had already been discovered in 1978, based on measurements of subtle deviations in the local magnetic field. These measurements had been made by geophysicist Glen Penfield of the Mexican oil company Pemex, but at the time no one connected them to the extinction of the dinosaurs. Thanks in part to the efforts of planetary researcher Alan Hildebrand, the crater finally received the attention it deserved in 1991.

The Chicxulub crater (named after a neighboring Mexican fishing village: the name means something like "devil's tail") is not visible from Earth's surface. It is buried under a half-mile-thick mass of limestone deposited over tens of millions of years. But in satellite measurements, the crater rim is recognizable as a minute disturbance in the relief of the (largely flat) landscape. Moreover, it is marked by several deep sinkholes (so-called cenotes) used by the ancient Maya as a source of drinking water. Precise gravity measurements, both on the surface and from space, also leave no doubt about the presence of the giant impact crater.

Over the years, more and more convincing evidence for the "dinosaur extinction impact" has come to light: shocked quartz crystals, for example, and traces of catastrophic tidal waves, especially in and around the Gulf of Mexico; tiny glass spheres (microtektites) in the thin K–Pg boundary clay layer formed when droplets of liquid rock solidified and fell back to Earth; and even fossil remains of fish that died shortly after the impact from the violence of a mega tsunami, in the Hell Creek Formation in North Dakota.

Bore holes in the so-called peak ring of the subsurface crater, conducted since 2016 from a large drilling platform off the coast of Yucatán, also support the theory that the mass extinction of 66 million years ago was the result of the impact. Increased iridium concentrations were found for the first time in the 910-yard drill core, exactly at the expected location. And microfossil measurements have confirmed that the wave of extinction at the K–Pg boundary coincided exactly with the creation of the crater.

But even though the evidence is piling up, some scientists are still not convinced. Geologist Gerta Keller from Princeton University continues to stick stubbornly to the volcano theory. Paleontologists have found indications that the dinosaurs were already in decline before the asteroid impact took place, and the climate reversal caused by the creation of the Deccan Traps would have eventually proved fatal for them. Smit and Keller have for years conducted heated discussions on these theories in the pages of scientific journals.

And maybe it is not a question of either/or—it could also be and/and. Extremely massive earthquakes resulting from the impact, which ripped violently through our planet's mantle, could have intensified volcanic activity on the Indian Plate. In that case, volcanic activity was the slow death knell, and the asteroid dealt the final blow to the dinosaurs indirectly. It cannot even be ruled out that there were multiple impacts: the oval Shiva crater on the ocean floor off the coast of India, hundreds of miles across, is believed by some to be the scar of a second catastrophic impact during the same period. A 15-mile crater in Ukraine is also about the same age. Either way, Earth was by no means a paradise 66 million years ago.

And what about earlier waves of extinction? Could these also have been caused by cosmic impacts? Paleontologists have identified four other major mass extinctions: at the Ordovician–Silurian boundary (about 445 million years ago), in the Upper Devonian (about 370 million years ago), at the Permian–Triassic boundary (252 million years ago, the worst catastrophe in the history of life on Earth), and at the Triassic–Jurassic boundary (201 million years ago). In none of these cases, however, has conclusive evidence been found for an extraterrestrial cause.

That has not prevented some scientists from speculating abundantly on this. American paleontologists David Raup and Jack Sepkoski once posited that some regularity can be found in the occurrence of mass extinctions, which becomes apparent if you include less catastrophic events.[3] Life on Earth would have had a tough time every 26 million years, the two scientists argued. Perhaps this is due to a periodic disruption of the Oort cloud: the shell of icy cometary nuclei surrounding the Sun at a great distance. Such a disruption (for example, due to the gravity of an as-yet-undiscovered dark companion of the Sun in a very wide, elliptical orbit) may cause a temporary increase in the number of comets colliding with Earth.

Speculation is rife, then, which is not surprising if you realize how little we know for certain about the geological history of our planet and the evolution of life. But the fact is that Earth is in the cosmic firing line, and occasionally suffers a direct hit. However, this is something the planet itself does not lose any sleep over; it can handle the odd collision. Fortunately, the impacts are also not nearly powerful enough to affect the orientation of Earth in space or its orbit around the Sun.

A collision with a 6- to 10-mile-wide asteroid is primarily disastrous for the biosphere—for fragile life on Earth. Dinosaurs, which had ruled the planet for tens of millions of years, did not see their sudden end coming. But we humans are well aware of the danger, and it would be completely irresponsible of us not to properly assess the risks, and not to think about ways to avert such a fate.

5
Extraterrestrial Impacts

On June 18, 1178, five English monks saw "the upper horn of the Moon split into two," after which a "flaming torch, hot coals and sparks" shot out from the celestial body.[1] At least, that was what chronicler Gervase of Canterbury described as happening. In 1976, geologist Jack B. Hartung posited the theory that the clerics had witnessed a cosmic impact on the Moon, during which the 13-mile-wide crater Giordano Bruno was formed. As seen from Earth, this crater lies on the hidden far side of the Moon and can only be viewed by Moon-orbiting spacecraft; however, it looks surprisingly "fresh," from what can be seen.

Today, Hartung's theory is no longer taken seriously: Giordano Bruno's age has been determined at around 4 million years. But the fact remains that the Moon is littered with large and small craters, and we now know that all of them were at some time caused by cosmic impacts. Lunar craters can be seen using even a relatively simple telescope. They were first recorded in November 1609 by Italian astronomer Galileo Galilei, who aimed his homemade telescope at the Moon. The largest lunar craters (also known as "walled plains") measure a few hundred miles across. Some, such as Copernicus and Tycho (many lunar craters are named after astronomers), have a remarkable "ray system"—bright streaks on the Moon's surface radiating away from the crater.

It took a long time before everyone was convinced that the craters on the Moon were caused by impacts. At the end of the seventeenth century, English physicist Robert Hooke wrote that they probably had a volcanic origin. And although geologists like Grove Gilbert and Ralph Baldwin later put forward strong arguments for the impact hypothesis, it was still hotly disputed into the middle of the last century. It was not until Eugene Shoemaker published his research on Meteor Crater that the last geologists were finally convinced. The lunar rocks that Apollo astronauts brought back with them to Earth removed any remaining doubts about their origin: the Moon is a cosmic battlefield.

The fact that there are far more impact craters on the Moon than on Earth is hardly surprising. The Moon has no atmosphere and exhibits hardly any geological activity. There is no erosion from water and wind or plate tectonics and mountain formation. As a result, the scars from the distant geological past are almost as fresh as when they were formed. For instance, the 58-mile-wide lunar crater Copernicus, some 80 million years old, would have eroded away long ago had it been located on Earth, just like the much larger Chicxulub crater, which is "only" 66 million years old. Our home planet looks so pristine simply because the traces of celestial violence have been erased quite thoroughly over time.

A look at other celestial bodies, however, reveals that the Moon is not the only one with so many impact craters. The innermost planet in the solar system, Mercury, looks just as battered. In the distant past, when the solar system was still filled with the debris of the process of planetary formation, it must also have rained stones, rocks, and asteroids on Earth. It is almost unbelievable that life could have developed amid that alien bombardment.

Far fewer craters have been discovered on Venus, the reason being that it has a very dense atmosphere, which means that smaller projectiles are slowed down or even disintegrate completely and vaporize. Venus also exhibits a lot of geological activity, which makes it much more like Earth: the few craters that do exist are relatively young.

Mars is somewhere in between. The planet's geological activity presumably died out long ago, and its atmosphere is much thinner than ours. Dust storms and wind erosion slowly but surely hid older craters from view, but large parts of the planet are just as battered as the surface of the Moon. Two large Martian craters, known as Gale and Jezero, are the home bases of the American Mars rovers Curiosity and Perseverance, respectively. The largest crater on Mars, measuring 290 miles in diameter, is called Huygens, after Dutch scientist Christiaan Huygens, who was the first person to observe surface details on Mars.

If you look for impact craters on Jupiter, Saturn, Uranus, and Neptune, you will be looking in vain: these giant planets are made up largely of gases, and they have no solid surface. But the dozens of moons that orbit around them often do have solid and crater-strewn surfaces. The same is true for the large asteroids Ceres and Vesta, which have now been visited by an unmanned space probe, and for large areas of the surface of the distant dwarf planet Pluto. Apparently, not a single celestial body in the solar system is safe from cosmic attack.

Planetary scientists are delighted with all these impact craters because they are a way of determining the ages of the surfaces of celestial bodies. A surface where there are few or no craters is clearly younger than a surface that is littered with them, as on the Moon. This makes it possible to reconstruct the geological history of a moon or planet to some degree. There are, however, some drawbacks to these crater counts: the impact frequency was not the same everywhere in the solar system, and, of course, we shouldn't let ourselves be misled by the numerous secondary craters formed by debris being hurled in all directions during a severe impact.

The study of craters on other celestial bodies—and especially on the Moon—has shown that things were much more violent in the distant past than they are today. Just under 4 billion years ago, there may even have been a real "bombardment," known as the Late Heavy Bombardment (where "late" refers to the fact that the solar system was already about 600 to 800 million years old at that time).[2] The possible cause: a disruption in the asteroid belt, due to the "migration" of the giant planet Jupiter to its current orbit. Among other things, the large impact basins on the Moon are thought to have formed during this period. These craters are in some cases over 600 miles in diameter and are now filled with solidified magma from the Moon's interior; they can even be seen with the naked eye as dark splotches on the Moon's surface.

The situation today may be less dramatic than it once was, but that does not mean that new craters are not still forming on the surface of our cosmic neighbors. Since 2005, telescopes on Earth have observed hundreds of flashes of light on the unlit part of the Moon. The brightest of these was recorded on March 17, 2013, and was probably caused by the impact of a large meteorite weighing around 90 pounds. Thanks to the Moon's lack of atmosphere, even small space rocks hit the surface at high speeds (10 to 40 miles per second), and a meteorite weighing just a few pounds can produce a crater about 10 yards across.

A careful comparison of old and new photos from space probes has indeed revealed brand-new impact craters on both the Moon and Mars. In 2020, planetary scientists published a catalogue containing at least a thousand of these new Martian craters (formed in recent decades), many of which are just a few yards in size.[3] Most were created by the impact of cosmic rocks not much larger than one or two feet—meteorites that would never reach the surface of Earth intact, thanks to our much denser atmosphere.

The surface of Mars is also teeming with smaller meteorites. Iron meteorites, which eventually decay on Earth from exposure to oxygen and moisture, remain preserved on the bone-dry Martian surface for many millions of years; at most, they get buried in windblown dust. Fifteen meteorites have already been found on Mars since 2005, mainly by the Opportunity and Curiosity rovers, even though they have only explored very small areas of the Red Planet.

Debris from impacts on other celestial bodies has also been found on Earth. Not surprisingly, the escape velocity of Mars (the speed required to escape the planet's gravitational field) is less than half that of Earth, and for the Moon it is even less than half as much again. So, if a heavy impact on the Moon or Mars ejects debris from the surface, some of it will end up in interplanetary space. Eventually, it may one day fall to Earth as meteorites.

It is possible to recognize a lunar or Martian meteorite by carefully examining the composition of the rock, or (in the case of a rock from Mars) by studying the composition of gas inclusions. Many hundreds of meteorites from the Moon and from Mars have already been identified using this method. An example of a very famous lunar meteorite, discovered in 2015 in Morocco, is NWA 10495, which is made up of several fragments weighing a total of 34 pounds. In 2019, one fragment of it, weighing 7 ounces, was sold by auction house

Christie's for $22,500—the private owner now has a real lunar rock on their mantelpiece. And remember the Nakhla meteorite that allegedly fatally struck an Egyptian street dog in 1911? That one is also known to have come from Mars.

The best-known Martian meteorite is undoubtedly ALH84001, discovered in 1984 at the foot of the Allan Hills in Antarctica. In 1996, NASA geologists believed they had found nanofossils of Martian organisms in the rock—a claim that has petered out completely over the course of time. However, the fact remains that the planets also occasionally bombard each other, and a fair amount of material must have been exchanged over the past billions of years.

The flashes of light observed on the Moon and the "new" impact craters found on Mars were all caused by relatively small projectiles. But in July 1994, astronomers witnessed a truly catastrophic cosmic impact on another celestial body when the 21 fragments of the disintegrating comet Shoemaker–Levy 9 penetrated the dense atmosphere of the giant planet Jupiter at high speed. This produced fireballs with temperatures of up to almost 45,000 degrees Fahrenheit and plumes of dislodged material thousands of miles high.

At first, many astronomers thought that the consequences of the impacts would be limited: the fragments were between 200 yards and 1.2 miles in size, while Jupiter itself is 870,000 miles in diameter. But the "scars" in Jupiter's atmosphere were visible from Earth even with a relatively small telescope: dark spots measuring many thousands of miles, created when material from deeper atmospheric layers was forced upward.

Since the dramatic spectacle of Shoemaker–Levy 9, more impacts have been observed on Jupiter, albeit less violent ones. The eighth such recorded, by amateur astronomers in Brazil and Germany, happened on the night of September 13–14, 2021—the very week I was writing the original version of this chapter, believe it or not. Clearly, Jupiter is hit by medium-sized projectiles much more often than Earth; not surprising, of course, given the planet's much stronger gravitational field. Exceptionally heavy impacts like that of Shoemaker–Levy 9 are estimated to occur once every few hundred years; a dark spot in Jupiter's atmosphere recorded in 1690 by Italian French astronomer Jean-Dominique Cassini may also have been caused by such a cometary impact.

It is quite remarkable that some 100 years ago, the danger of cosmic impacts was barely recognized and certainly not taken seriously by most scientists. Today, we know that the evolution of life on Earth has been profoundly affected by such impacts, we have identified the traces of extraterrestrial violence all around us, and we have witnessed a spectacular collision of a comet on another planet. The key question now is: How much danger is Earth actually in? Are there in fact deadly projectiles on a collision course with our home world?

6
Near Misses

Here is a date for you to put in your diary: on April 13, 2029, a speck of light will be visible in the sky, moving slowly with respect to the other stars. Bright enough to be seen with the naked eye (particularly from Europe), it is the nearly 400-yard-wide asteroid Apophis, flying close to Earth—at a distance of "only" 19,600 miles. Back in late 2004, it even briefly seemed that Apophis might collide with Earth 25 years later on that April thirteenth (a Friday, by the way!).

Apophis was discovered in the summer of 2004, and passed Earth on December 21 of that year at a safe distance of 8.9 million miles. But during this encounter, Earth's gravity slightly deflected the small celestial body's orbit. As a result, there seemed to be a 2.7 percent chance of a catastrophic collision in 2029. This worrying news was overshadowed by reports of a real natural disaster: the earthquake and resulting tsunami that claimed over 200,000 lives in Indonesia, India, and Thailand on Boxing Day 2004. Yet the impact of a 400-yard-wide space rock would be even more disastrous.

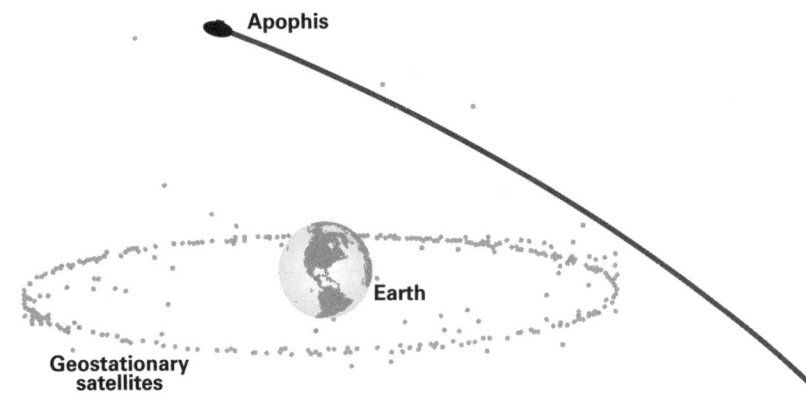

Figure 6.1
On April 13, 2029, the 400-yard-wide Earth-grazing asteroid Apophis will have a near miss with Earth, passing just outside the orbits of geo-stationary satellites. Source: NASA/JPL.

Since the coronavirus pandemic, we're all aware that keeping our distance can be of vital importance. But astronomers have a very different idea of the term "near." Even if an asteroid passes Earth at a distance of a few million miles, they call it a near miss. Such an intrusive celestial body is called a near-Earth asteroid (NEA), a near-Earth object (NEO), or an Earth-grazing asteroid. In 2029, Apophis will certainly live up to those names; it is not without reason that the asteroid was given its Greek name, taken from the Egyptian god Apepi, the all-destroying adversary of the Sun god Ra.

To find out for sure whether we really do need to worry, a detailed study was made of Apophis in January 2013, when it was again more or less close to Earth. The use of radar observations allowed its orbit to be determined much more precisely. The reassuring message was: no impact in 2029 (although 19,600 miles is really, worryingly close); no impact on the next close approach in 2036; and only a 1 in 250,000 chance of a collision with Earth in April 2068. For now, we can all breathe a sigh of relief.

How is it possible for an asteroid to get so close to Earth when most of these small, rocky celestial bodies orbit the Sun between the orbits of Mars and Jupiter? This is indeed the case for the vast majority of asteroids: they pose absolutely no danger to Earth. But due to their mutual gravitational influence, the orbits of asteroids can occasionally become seriously disrupted. Thousands of such asteroids are known to have elongated elliptical orbits that are often also tilted with respect to the orbits of the planets. These asteroids are able to traverse the inner parts of the solar system, posing the risk of a collision with our planet.

On August 13, 1898, German astronomer Carl Gustav Witt was the first to discover an asteroid venturing within the orbit of Mars. Eros, as the minor planet was named (after the Greek god of love), poses no concrete threat at present. But such an anomalous orbit is not stable, and calculations show that the elongated orbit of the celestial body could also start crossing Earth's orbit within 2 million years. Not a pleasant prospect given its dimensions of 21 by 10 by 7 miles.

In 1932, German astronomer Karl Reinmuth first discovered an asteroid (called Apollo) whose orbit overlaps that of Earth. And in 1949, German American astronomer Walter Baade observed a rock about a mile in size in an extremely elongated orbit crossing the orbits of Mars, Earth, Venus, and Mercury. This was the first discovery of an asteroid that could come closer to the Sun than Mercury. It was quite appropriately named Icarus, after the reckless youth from Greek mythology who tried to fly too close to the Sun with wings made of wax. On June 16, 2015, Icarus passed Earth at a safe distance of about 5 million miles—about 20 times the distance between Earth and the Moon.

There are currently more than 12,000 known Earth-crossing asteroids, and every now and then one passes at a relatively short distance. One example is 1998 OR_2, a lump of rock 1.2 miles across that approached our planet on April 29, 2020, at a distance of 3.9 million miles. Another is 2001 FO_{32}, which sped past on March 21, 2021, at 77,000 miles per hour and closer than 1.2 million miles—three times closer than 1998 OR_2, but luckily also only one-third its size. Smaller objects are found much more frequently in the vicinity of Earth, for the simple reason that they are much more numerous. To give you an idea: on September 16, 2021, the 6.5-yard-wide asteroid 2021 RF_{16} passed at a distance of just 186,000 miles, within the Moon's orbit.

Of course, not all Earth-crossing asteroids pose the same danger to our home planet. Astronomers speak of a "potentially hazardous asteroid" (PHA) if it is larger than some 150 yards and if the shortest distance between the asteroid's orbit and Earth's orbit is less than 4.7 million miles (7.5 million kilometers). It is nonetheless worth keeping a close eye on such celestial bodies because they could well cause a truly catastrophic impact in the future. As of November 2022, there were 2,304 such PHAs known, 153 of which are over 0.6 miles in diameter. It is estimated that, on average, Earth is struck by such a PHA once every 10,000 years.

As mentioned before, smaller NEOs are much more numerous, so they are also more likely to collide with our planet. But the consequences are, of course, far less severe—think of the 2013 Chelyabinsk impact. At the turn of the century, astronomers devised the Torino scale (presented at a conference in Turin, Italy, hence the name) to indicate the severity of an impact threat on a scale of 0 to 10. This scale takes into account the probability of an object actually colliding with Earth, as well as the size of the projectile. NASA's Jet Propulsion Laboratory keeps a record of potential impacts over the next few hundred years (cneos.jpl.nasa.gov/sentry/), but in all these cases the objects are relatively small and the impact probability is so negligible that the Torino value never exceeds 0.

The largest asteroid on this Sentry list is Bennu, with a diameter of about 550 yards. There is a less than 1 in 1,000 chance that Bennu will collide with Earth sometime between 2178 and 2290; the date with the highest impact risk is September 24, 2182. But that is a long time in the future and the probability is very low, so nobody is particularly worried about it. Moreover, the orbits of asteroids are constantly subject to gravitational disturbances, which cannot always be predicted accurately in advance. In the summer of 2150, for example, the 0.8-mile-wide near-Earth object 1950 DA will pass the much larger asteroid Diana, which has a diameter of 75 miles, at a distance of 280,000 miles. Diana's gravity will deflect 1950 DA's orbit, but no one can predict exactly how strongly.

It is not only asteroids that come close to Earth; comets also sometimes fail to keep a safe distance. And there's another very different problem with comets: you generally don't see them coming far in advance. Most comets trace extremely elongated orbits around the Sun and are invisible from Earth for tens or even hundreds of thousands of years. Such a "new" comet is often not discovered until it has more or less arrived near the Sun again, at a distance of at most half a billion miles. Only then can its orbit be determined and it become clear whether or not the comet is heading for an unfortunate encounter with Earth. So, while you can often foresee potential collisions with asteroids many decades in advance, this is almost never possible with comets: they really do launch a surprise attack.

Near-Earth objects also include short-period comets, such as comet Lexell. This small, inconspicuous lump of ice orbits the Sun in an elongated orbit once every 5.6 years. It was discovered in the summer of 1770 when it skimmed past Earth at a distance of 14 million miles: the closest comet approach ever known to have happened. Incidentally, Earth has passed through the extremely tenuous tail of a comet more than once, such as in 1910 with the famous comet Halley. However, a cometary tail contains only gas atoms and microscopic dust particles, so it is not a cause for concern.

Similar to long-period comets—at least in terms of their unannounced appearance—are interstellar objects: small celestial bodies that do not orbit the Sun but traverse the space between the stars, inadvertently passing right through our solar system. This kind of interstellar object was first discovered in October 2017, in the form of the enigmatic celestial body 'Oumuamua, which was even briefly thought by some to be a spaceship from an alien civilization. Two years later, the newly discovered comet Borisov also turned out to be from outside the solar system. Even such cosmic passers-by could, of course, make a once-in-a-lifetime appearance as NEOs—or, if things get really bad, as interstellar dodgeballs.

Astronomers will never have a complete inventory of all near-Earth objects or potentially hazardous asteroids that could be on a collision course with Earth. This is partly because new objects, such as comets and interstellar objects, that fly into our solar system from the shadowy depths of the universe are constantly being discovered. Moreover, most asteroids are simply too small for us to discover them from Earth far in advance; the Chelyabinsk impact, for instance, was completely unexpected, and, unfortunately, Earth could be hit tomorrow (or today!) by a death-and-destruction-bringing projectile 100 yards across that nobody sees coming.

A third reason why the future will always remain somewhat uncertain is that it is impossible to accurately predict the behavior of NEOs in the distant future. Orbital calculations of asteroids today take into account the gravity of the Sun, the planets, their large moons, and many hundreds of other asteroids, but these calculations are never 100 percent exact, if only because the masses of all these celestial bodies are not known precisely enough. Moreover, the so-called butterfly effect (known from chaos theory) also applies to the solar system: a minute perturbation can eventually lead to extremely large deviations, just as in the metaphor of the wingbeat of a butterfly in Africa eventually causing a hurricane in the Americas.

One of the hard-to-predict effects in the orbital calculations of small, rotating asteroids is the Yarkovsky effect, named after the Polish Russian engineer Ivan Osipovich Yarkovsky who was the first to describe it around 1900. As an asteroid orbits the Sun, its dayside is heated up, but as the asteroid rotates, the hottest part of its surface is no longer precisely facing the Sun. The thermal energy from the asteroid is therefore emitted not in the same direction from which the sunlight originates but in a somewhat different direction. This causes a minute rocket effect, which changes the asteroid's orbit slightly over the course of many years.

The Yarkovsky effect can have major consequences over time, but it can also have benefits. At the beginning of this chapter, you read about the NEO Apophis. This will miss Earth by a hair's breadth on Friday April 13, 2029, but, as mentioned, measurements in 2013 showed that there was still a 1 in 250,000 chance of Apophis colliding with Earth in April 2068. This risk assessment has since been revised. Comparing the 2013 measurements with new observations made since 2020 has given a much more accurate estimate of the magnitude of the Yarkovsky effect for this asteroid. This has also made it possible to make much more precise statements about the future, and astronomers now know for sure that Apophis does not pose a real impact threat to Earth in 2068 (nor for centuries to come). As a result, this potential troublemaker is also no longer on the Sentry list of NASA's Jet Propulsion Laboratory. Knowing that, we can sleep just a little more peacefully.

7
Open Season

On November 13, 2020 (yes, a Friday!), a space rock some 8 yards in diameter flew past Earth, just 230 miles above the surface—lower than the International Space Station. The good news: the projectile (known as 2020 VT$_4$) was discovered by a telescope in Hawaii. The bad news: that only happened 15 hours after the asteroid had passed its closest distance from Earth, when it was already hurtling away again.

Every night, telescopes search the sky for fast-moving specks of light: small asteroids that could at some point in the future collide with Earth. The more sensitive the detection equipment becomes, the larger the number of asteroids that come into view. The calendar year 2020 has taken the crown so far, with a total of 2,958 newly discovered near-Earth objects. But then again, you can only see small objects when they are already close to Earth; if they approach our planet on the dayside you can't see them at all; and in any event it is impossible to look continuously in every direction. Therefore, the dis-covery sometimes takes place after the actual passage, as happened with 2020 VT$_4$. Or not at all, as was the case with the Chelyabinsk impact in 2013.

The hunt for asteroids started even before the first ones were found. Inspired by the discovery of the planet Uranus in 1781, Hungarian astronomer Franz Xaver von Zach, together with some 20 European colleagues, started a huge search for an unknown planet expected to exist in the surprisingly wide, empty area between the orbits of Mars and Jupiter. Asteroid number 1, Ceres, was discovered by chance on January 1, 1801, by Giuseppe Piazzi, but numbers 2, 3, and 4 (Pallas, Juno, and Vesta) were discovered thanks to the detective work of Zach's "celestial police," as he called his search team. After this, the number of known asteroids increased rapidly, from 10 in 1849 to 100 in 1868 to 1,000 in 1921.

Dutch American astronomer Tom Gehrels launched a new search program in 1960, in collaboration with his Leiden Observatory colleagues Kees van Houten and Ingrid van Houten-Groeneveld. Gehrels made hundreds of images of the night sky with the 48-inch Schmidt telescope at the Palomar Observatory in California: always two of the same area of sky, but at different points in time. The van Houtens used a so-called blink comparator to compare the images and hunt for moving specks of light. This instrument allows you to switch quickly between the two photos; the stars remain stationary, but an asteroid immediately stands out because it jumps back and forth. Between 1960 and 1977, the Palomar–Leiden Survey yielded no fewer than 4,637 new asteroids, including a number of NEOs.

The success of the Palomar–Leiden Survey inspired Gehrels to go further. In 1984, together with colleague Bob McMillan from the University of Arizona, he started the Spacewatch Project, which focuses entirely on detecting asteroids that can come dangerously close to Earth. Initially, they used a 35-inch telescope at the Kitt Peak National Observatory southwest of Tucson, Arizona, but in 2000 they switched to a much more sensitive instrument with a primary mirror 90 inches in diameter. This is the world's largest telescope used exclusively for searching for and studying NEOs.

Gehrels died in 2011, but the 90-inch Spacewatch telescope is still going strong. In terms of new discoveries, Spacewatch remains the undisputed record holder: as of 2022, the number of discoveries stood at over 179,000! Incidentally, the smaller 35-inch telescope is also still being used, while two more sensitive instruments on Kitt Peak—the 90-inch Bok Telescope and the 157-inch Mayall Telescope—can be used to study newly discovered asteroids in more detail. Today, sensitive electronic cameras have of course replaced photographic glass plates, and there is no longer any need to search by eye for moving specks of light; computer algorithms have been developed to do that.

It did not take long for other astronomers to follow Spacewatch's example. US astronomer Ted Bowell's LONEOS (Lowell Observatory Near-Earth-Object Search) project began hunting for near-Earth objects in 1993 using a medium-sized telescope at the Lowell Observatory in Flagstaff, also in Arizona. Two years later, in 1995, NASA's NEAT (Near-Earth Asteroid Tracking) program began, led by US astronomer Eleanor Helin. NEAT used two telescopes, one near the summit of the Haleakalā volcano on Maui, Hawaii, the other on Palomar Mountain in California. LONEOS was operational until 2008; in its 15 years of service, the project discovered 22,724 new asteroids. NEAT ended in 2007, with a score of 43,208 new discoveries.

Since 1998, two other US projects have been mapping most of the flying debris in the inner solar system: the previously mentioned Catalina Sky Survey that uses several telescopes at the Steward Observatory's Catalina Station in the Santa Catalina Mountains northeast of Tucson, and the Lincoln Near-Earth Asteroid Research (LINEAR) project near Socorro, New Mexico. They were joined in 2015 by the ATLAS (Asteroid Terrestrial-impact Last Alert System) project in Hawaii. Currently, Catalina is the more prolific of the three: almost half of all known NEOs have been discovered as part of this survey, with many hundreds more being added every year. In many cases, these are objects of at most a few yards in size that are not seen until they are already close to Earth. The Catalina Sky Survey discovered the two small asteroids that came down over Sudan and Botswana, the meteorite remains of which were found by Peter Jenniskens and his colleagues.

If you want to know everything that is coming at you, you obviously have to be able to look all around you in every direction. It has already been noted that this is difficult to do on the dayside of Earth, where you are looking in the direction of the Sun. But unfortunately there is yet another blind spot in the hunt for NEOs: most active search programs are carried out from Earth's northern hemisphere. Similar surveys in the south (notably the Anglo-Australian Near-Earth Asteroid Survey and the Siding Spring Survey) were terminated many years ago due to the loss of government funding, and at present, there is only one southern-hemisphere survey in operation, since early 2022. In short, the inventory is by no means complete yet.

Nonetheless, thousands of potentially hazardous asteroids (PHAs)—objects of at least 150 yards in size that can come dangerously close to Earth—have already been found. The largest is the unnamed asteroid 53319, discovered in 1999 by LINEAR. It can approach Earth's orbit to within 2.2 million miles and has a diameter of 4.5 miles—not much smaller than the projectile that wiped out the dinosaurs. Another remarkable PHA is number 85713: the smallest distance between this celestial body's orbit and Earth's orbit is just 217,000 miles—less than the average distance to the Moon. A catastrophic collision in the distant future seems almost inevitable, but that could well be millions of years away.

Detecting celestial objects is one thing, but you also need to learn more about the newly discovered objects. Large telescopes are used for this, as well as radar techniques. By sending radio waves to an asteroid and collecting the echoes, astronomers can learn a lot about its

size, shape, rotation characteristics, and—very impor-
tantly—its exact position in space. This kind of radar
research has been taking place since 1968; in the summer of 2021, it was used for the thousandth time to
study a small asteroid.

The newest search programs are fully automated. The
two 70-inch Pan-STARRS (Panoramic Survey Telescope
and Rapid Response System) telescopes at the Haleakalā
Observatory in Hawaii are equipped with gigantic 1.4
gigapixel digital cameras that have a field of view six
times the apparent diameter of the full Moon. The
cameras produce 10 terabytes of observational data
every night. Powerful supercomputers search the raw
data not only for NEOs but also for distant supernova
explosions—with no human intervention at all.

The same applies to the ATLAS project. ATLAS oper-
ates two 20-inch telescopes in Hawaii, but despite their
smaller size, they have an even larger field of view than
Pan-STARRS's telescopes. The aim is primarily to detect
small objects that will actually collide with Earth, so
that responsive actions can be taken at the last minute—
hence the term "last alert." An impactor of 50 to 100
yards in diameter can be found by ATLAS a few days
before the impact—hopefully long enough to evacuate
the target area. ATLAS has recently been expanded to
include telescopes in the southern hemisphere, in Australia and Chile.

It all sounds very logical: if you know that you are
going to be under fire, you need to be constantly on

guard to see a new attack coming. The one thing you don't want is to be surprised by it, as the dinosaurs were 66 million years ago. The idea of such a Spaceguard was suggested back in 1973 by science fiction writer Arthur C. Clarke in his book *Rendezvous with Rama*. "Spaceguard" is now also used as the collective name for all these global search and survey projects.

The vast majority of these projects, as you may have noticed, are American, funded mainly by NASA. In 1998, the space agency was mandated by the US Congress to detect at least 90 percent of the estimated 1,000 potentially hazardous asteroids larger than 0.6 miles in size and map their orbits. That goal had been achieved by 2011. Work is now underway to track all PHAs larger than 150 yards; there should be some 15,000 of them. NASA's Planetary Defense Coordination Office has an annual budget of $150 million.

The hunt for near-Earth objects is not only conducted from the ground. Astronomers also keep a close eye on our planet's surroundings from space. NASA's WISE (Wide-field Infrared Survey Explorer) space telescope, for instance, has spotted no fewer than 160,000 asteroids, more than 35,000 of which had never been seen before. WISE was launched in late 2009 and served as an infrared (deep-cooled) space observatory until February 2011. After a kind of hibernation period, the satellite was reactivated as an asteroid hunter in September 2013, under the name NEOWISE, with NEO standing for Near-Earth Object. If you think the name looks familiar to you, you may be right: in early 2020, the satellite discovered the comet later named Neowise, which was visible to the naked eye in the night sky that summer.

A lot of hard work is being done on a successor, named NEO Surveyor. It will have a larger telescope and more sensitive detection equipment, and will also be stationed at a considerable distance from Earth: almost six times as far as the Moon, in the direction of the Sun. By monitoring the area around Earth from that position, NEO Surveyor will also be able to detect objects approaching Earth on the dayside—something that is virtually impossible from Earth's surface or from an orbit around Earth. NEO Surveyor is set to launch in 2027 and is expected to detect at least 100,000 new asteroids, including undoubtedly a good number of PHAs.

Well before the launch of NEO Surveyor, the Vera C. Rubin Observatory will be inaugurated on the Cerro Pachón mountaintop in northern Chile—probably in early 2025. The new observatory's colossal 330-inch telescope is equipped with a 3,200-megapixel camera, the most sensitive digital camera ever built. This camera will capture the entire starry sky over Chile every three to four nights, making it possible to detect asteroids 100 yards in size at distances of several hundred million miles. This new showpiece of astronomy is expected to be able to detect, in just one year, thousands of objects comparable to the 50-yard space rock that razed 1,200 square miles of Siberian pine forest in 1908.

Asteroid hunters cannot wait for their new toys. But at the same time, this deluge of discoveries also confronts us with an important question: If you know exactly what is flying around up there and what might be on a collision course with our planet, what do you do with that information? Suppose we actually do discover a 100-yard near-Earth object that is on course to crash down somewhere on Earth in the not-too-distant future—what can we do about it?

8
Defending the Planet

It should be clear by now that our solar system is a celestial shooting gallery, and Earth is in the firing line. Projectiles are flying all around us, and the question is not *if* we will suffer a direct hit but *when*. British physicist Stephen Hawking even believed that a cosmic impact represents the greatest threat to humanity, far greater in any event than a global pandemic or a terrestrial natural disaster. A vaccination campaign will not help against an oncoming comet or asteroid, and so the big question is: What can be done about it? How can we defend the planet and humanity against extraterrestrial violence?

There's one thing we can be sure of: it will never be completely safe here on Earth. Every day, meteorites land on Earth's surface, and although most of them end up in the ocean, you could be unlucky enough to have one land on your head—or on some other part of your body, like Ann Hodges. Larger cosmic rocks, like the space rock that exploded above Chelyabinsk in 2013, are something you simply won't see coming. So, yes, there are almost certain to be casualties from falling meteorites at some point in the future, and there is nothing we can do about it. Fortunately, the consequences of such an impact are usually local and limited; earthquakes, volcanic eruptions, and tsunamis claim far more victims.

Unfortunately, we humans are also powerless against the rare giant projectiles that are many miles in diameter. Unlike the dinosaurs, we might well see the approach of a 6-mile-wide killer asteroid, like the one that collided with Earth 66 million years ago, but stopping it or deflecting its course is out of the question: it would be like trying to stop an oncoming truck by throwing ping-pong balls at it. Fortunately, we have now discovered all near-Earth objects larger than a few miles (and there are none on a collision course), but astronomers could very well discover an enormous comet next week that will crash into Earth in a few years' time. And there's nothing we could do to stop it.

If we want to protect ourselves from cosmic impacts, we need to focus on the medium-sized objects, ranging from, say, 100 yards to about half a mile. These are relatively numerous, and they can easily cause many tens of millions of casualties. Earth is hit by a 400-yard asteroid on average once every 100,000 years. If the collision occurs in Europe, a country like France will disappear completely from the map, and the whole continent will be one unimaginable disaster area. Such an impact is, in theory, preventable, so we would be crazy not to explore the possibilities of doing just that.

ASTEROID DIAMETER (METERS)

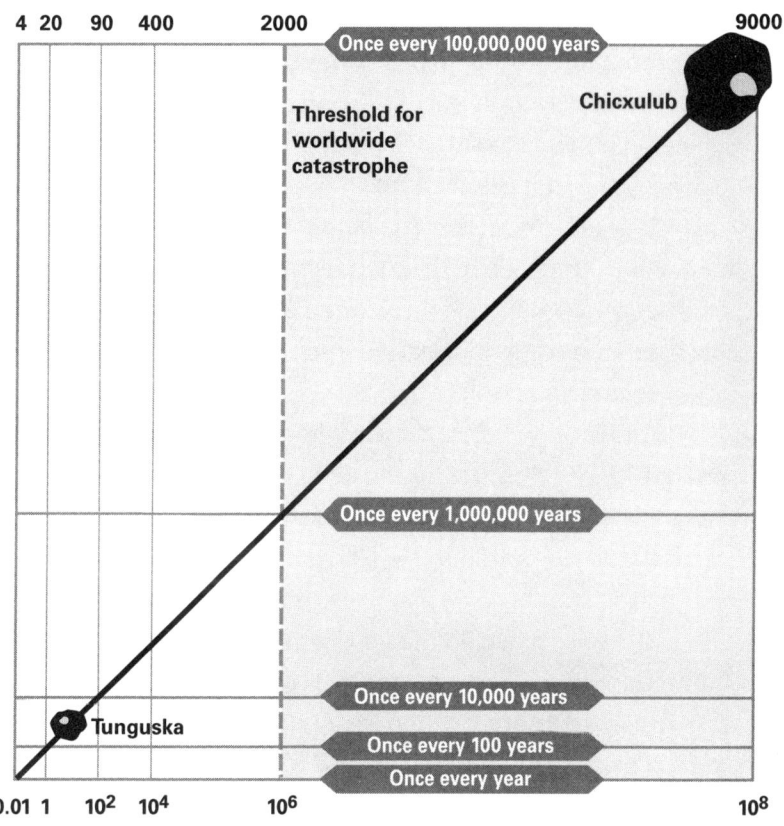

IMPACT ENERGY (MEGATONS TNT)

Figure 8.1
The relation between the size of a cosmic projectile, the associated impact energy, and the average impact frequency. Source: NASA.

That's what Dutch astrophysicist Piet Hut of the Institute for Advanced Study in Princeton, New Jersey, thought too. A few years after the 1998 Hollywood blockbusters *Deep Impact* and *Armageddon* brought the general public face-to-face with the possibility of an impact, Hut organized a workshop on the possibilities of averting such doomsday scenarios. A year later, in October 2002, together with a fellow astronomer and two former astronauts, he founded the B612 Foundation—a private nonprofit foundation that aims to investigate how to deflect approaching celestial bodies.

B612 was the name of the asteroid on which the Little Prince lived in the famous 1943 children's book by Antoine de Saint-Exupéry. (The hexadecimal code B612 stands for the number 46610; that's why the asteroid with this number was later also named Bésixdouze— B-6-12 in French.) Ten years ago, the foundation had ambitious plans for a satellite of its own, called Sentinel, to search for potentially dangerous asteroids, but the project was cancelled due to lack of funds. Even so, the B612 Foundation is still one of the leading advocates of serious research into planetary defense techniques.

Meanwhile, government organizations such as NASA and the European Space Agency (ESA) are also not sitting idly by. As mentioned earlier, NASA has its own Planetary Defense Coordination Office, while ESA is pouring money into the NEOShield program. The US National Science and Technology Council has developed its own National Near-Earth Object Preparedness

Strategy, and even within the United Nations Committee on the Peaceful Uses of Outer Space (COPUOS) there is an action team dealing with the danger of cosmic impacts. Besides its own International Asteroid Warning Network, the UN now also has a Space Mission Planning Advisory Group. In short: at the very least, a lot of meetings are being held on how to protect humanity from attacks from the cosmos.

There's one thing we know for certain: blowing an asteroid up with an atomic bomb, as happened in *Armageddon*, is not a smart idea. It is an option that was put forward seriously a long time ago by Edward Teller (the spiritual father of the hydrogen bomb), but it simply wouldn't help. The numerous fragments created in such an explosion would still be moving through the solar system in more or less the same direction and at the original high speed. As a result, Earth would then have to endure not one big impact but a whole series of smaller ones, with all the attendant consequences.

A more practical solution would be to slightly deflect the approaching celestial body so that it passes close to Earth rather than colliding with it. Particularly if you can see the impact coming many years in advance, a small nudge can be enough to avert disaster. When astronomers discovered the NEO Apophis, which for a while looked as if it would wreak havoc on Earth in 2029, they were already calculating that a minimal change in speed of just a few micrometers per second would be enough to prevent that anticipated catastrophe.

NASA has gained experience with targeting small celestial bodies: in 2005, the *Deep Impact* space probe fired an 820-pound copper projectile at the nucleus of comet Tempel 1, with the goal of studying the structure and composition of the lump of ice. That was not enough to noticeably change the hefty comet's course, but at least we do have the technology.

At Lawrence Livermore National Laboratory, the HAMMER project is on the drawing board. HAMMER (Hyper-velocity Asteroid Mitigation Mission for Emergency Response) is a celestial battering ram, 10 yards long and weighing almost 9 tons, that can be fired at high velocity at a small near-Earth object. With a 10-year warning period, it could deflect a 100-yard-wide object enough to prevent an impact. If something larger is speeding toward Earth, you just send out 10 or 20 HAMMERs. Or 50, or 100. Admittedly, that is a hugely expensive proposition, but if it means you can save 100 million lives, cost is obviously a secondary consideration.

Incidentally, there is a cheaper way to nudge a small asteroid out of its original orbit: just place a giant rocket motor on its surface. If a small rocket motor can transport a launcher into space, a big one should let you accelerate or decelerate an entire NEO at least a tiny bit. As the raw material for the rocket fuel, you should preferably use the material of the asteroid itself: hydrogen can be extracted from ice, and oxygen from rock. Or you simply catapult material from the object into space at high speed—thanks to Newton's third law (every action produces an equal and opposite reaction), that too results in a kind of rocket effect in the opposite direction.

Something else that also works is heating a small area on one side of the asteroid so much that the surface material evaporates and jets off into space. The effect is the same as that of a rocket engine on the surface: gas is blasted away in one direction, propelling the asteroid a tiny bit in the other direction. If you can set a piece of paper or a shoelace on fire using a magnifying glass, you can also focus sunlight on the surface of an asteroid using a large swarm of satellites equipped with gigantic lenses. An entire fleet of laser cannons is also an option, of course, or a nuclear explosion at a small distance from the celestial projectile.

There is a problem, though: the effectiveness of all these techniques depends heavily on the structure of the asteroid in question. A solid lump of rock or a gigantic lump of iron a few hundred yards in size is likely to react completely differently from a flying cloud of gravel made up of numerous individual rocks held together by gravity (a so-called rubble pile). Another suggestion is to wrap an approaching NEO in thin, reflective foil (the American wrapping artist Christo would have loved it!)—this would either strengthen or weaken the Yarkovsky effect mentioned in chapter 6, which would also affect the orbit of the celestial body. Giving it a once-over with a can of spray paint is, of course, another possibility.

A less invasive option is the idea of the gravitational tractor, developed by former astronaut Ed Lu (cofounder of the B612 Foundation) and his colleague Stan Love. A gravitational tractor device, such as a large, heavy space probe, would fly alongside the near-Earth object for an extended period (years to decades) and slowly drag it away from its collision course. The probe would have to have its rocket engine on the whole time, otherwise it would itself be attracted by the gravity of the celestial body. With a bit of careful maneuvering, you could pull a killer asteroid into a safe orbit. At least, if you have enough time, because this method would take many years to be effective.

Optimistic space engineers even think that the impact of a much larger object, a few miles in diameter, could be prevented by a game of cosmic billiards. One of the techniques described above could be used to deflect a smaller asteroid very precisely, causing it to fly past the larger killer asteroid at a relatively short distance. The mutual gravitational force would have a small but significant effect on the movements of both celestial bodies, preventing a catastrophic impact.

It all sounds fantastic, but there are also some complex political obstacles in the whole idea of planetary defense. Suppose a relatively small near-Earth object is speeding toward our planet, threatening to wipe the US city of Dallas (population over a million) off the map. Will Russia and China be willing to help pay for a "rescue mission"? Do Americans have money to spare for the preservation of Chengdu? Do people in Europe care about the possible fate of Zimbabwe?

American astronomer Carl Sagan foresaw yet another problem: if a country has the possibility of deflecting a small asteroid such that it passes close to Earth, the same technology can also be used to bring the asteroid down on an enemy. On this basis, the utopian concept of planetary defense could also turn into a celestial version of the Cold War—or worse. These are exactly the kinds of issues that are on the agenda of the UN special committee dealing with the threat of cosmic impacts. For the time being, however, any form of consensus is still a long way off.

Nonetheless, something has to be done. If you are in the firing line, you have to protect and defend yourself as best you can, otherwise a death sentence is a foregone conclusion. Doing nothing is simply not an option. Identifying the danger, studying all the conceivable countermeasures, and being ready to act when necessary—we owe this not only to our descendants but to humanity as a whole. As with fighting the coronavirus pandemic and the climate crisis, the urgency of the problem will likely only sink in when the need arises. Hopefully, it won't be too late by then.

9
Rendezvous with the Aggressor

It was a kamikaze mission: on September 26, 2022, the American spacecraft *DART* slammed into a 175-yard-wide cosmic rock at high speed. The 1,100-pound probe, with its gigantic solar panels, did not survive the impact. But if all went according to plan, the asteroid would be nudged very slightly off course. Indeed, measurements from Earth showed that this is what happened. It was the first time such an experiment had been carried out, so no one knew in advance exactly what to expect.

The *DART* (Double Asteroid Redirection Test) spacecraft was launched on November 24, 2021. After a nine-month journey, it reached the 850-yard-wide near-Earth object Didymos, an asteroid that was discovered in 1996 by the Spacewatch program. In 2003, astronomers learned that Didymos was accompanied by a smaller body, some 175 yards across and around half a mile away, that was later named Dimorphos. This tiny moon was *DART's* target. The energy of the collision reduced Dimorphos's original orbital period (11.9 hours) by 32 minutes.

This was not the end of the experiment. In October 2024, the European Space Agency launched the *Hera* space probe. The probe should arrive in an orbit around Didymos and its small moon in early 2026, where it will carry out research on the crater that may have been formed by *DART*. *Hera* is also going to make a detailed study of the composition of the two celestial bodies. The aim is to give us an idea of what is literally hanging over our heads.

Space research on asteroids and comets—the "small stuff" in the solar system—is not just interesting for those who want to prepare for future cosmic impacts. Planetary researchers realized long ago that these celestial bodies hold the key to the origin of the solar system. Unlike the larger planets, asteroids and comets are not "differentiated" (the process by which heavy elements like iron and nickel sink to the core of a celestial body), so, in terms of chemical composition, their surface still closely mirrors the material from which the planets coalesced. Moreover, the current distribution of celestial lumps of rock and ice tells us something about the dynamic evolution of the solar system.

The first time an asteroid was visited by a space probe was in October 1991, when NASA's Jupiter explorer *Galileo* flew past Gaspra—an elongated clump of rock measuring 11 by 6 miles—at a distance of approximately 1,000 miles. Just under two years later, *Galileo* also passed the much larger asteroid Ida (measuring about 35 by 12 miles). Both celestial bodies showed signs of numerous small impact craters, and—at least from the outside—seemed very similar to the two small moons of the planet Mars, which are possibly captured asteroids. A moon called Dactyl, with a diameter of about 1 mile, was discovered near Ida—the first confirmed moon of an asteroid.

It is one thing to fly past an asteroid, but landing on one is an entirely different matter. The first successful landing was on February 12, 2001, when the American space probe *NEAR Shoemaker* landed on the surface of NEO Eros, which is about half the size of Ida. *NEAR Shoemaker* (NEAR stands for Near Earth Asteroid Rendezvous; the space probe is named after geologist and impact expert Eugene Shoemaker) first orbited Eros, examining the elongated rock from all sides. After the semisoft landing, the probe remained in operation for over two weeks.

Incidentally, in the summer of 1997, *NEAR Shoemaker* had flown past the asteroid Mathilde, which is over 30 miles in size, at a distance of 745 miles. It turned out to be exceptionally dark—as black as pitch—while its low density indicated that it is probably what is known as a rubble pile: a loose collection of smaller rocks and stones held together by gravity.

If a space probe crosses the asteroid belt on its way to its destination, you might as well get it to take a look at one of these miniature planets. Following the successful fly-bys by *Galileo* and *NEAR Shoemaker*, this is something that has been done more often. *Deep Space*

1 paid a visit in 1999 to the asteroid Braille; in early 2000, the Saturn probe *Cassini* flew past Masursky (at a very considerable distance); asteroid Annefrank received a visit from comet explorer *Stardust* in November 2002; and the European space probe *Rosetta* even flew past two asteroids: Šteins—a lump of rock around 3 miles across—in 2008, and the 60-mile-wide Lutetia in 2010. And then there was the Chinese lunar probe *Chang'e 2*, which, after studying the Moon, flew right past the small but potentially dangerous NEO Toutatis in December 2012. All this is bonus material for other areas of space research, which has provided a breadth of insights into the enormous variety of the solar system's tiny wanderers.

The most detailed study carried out on asteroids so far was done by NASA's space probe *Dawn*. Launched in September 2007, it has orbited the two largest inhabitants of the asteroid belt: Vesta (around 356 miles in diameter) and Ceres (585 miles in diameter). It was the first time a single space probe was put into orbit around two different celestial bodies (if you exclude Earth). This was made possible by *Dawn*'s flexible ion propulsion system—a technology that has since been applied more often.

A thorough study was made of Vesta between July 2011 and September 2012. The study showed that Vesta has been severely battered by a cosmic collision, which created the giant impact basin Rheasilvia, at the asteroid's south pole. Vesta is differentiated; as far as we know, it is the only remaining rocky "protoplanet" from the period when the solar system was formed. The dwarf planet Ceres proved to be interesting for an entirely different reason: it contains a lot of ice; there may be an underground ocean, and its surface shows signs of "cryovolcanism," as evidenced by the 2.5-mile-high ice volcano Ahuna Mons.

Obviously, it is nice to study an asteroid up close through the instruments aboard a space probe, but researchers would far rather have samples from a celestial body under a microscope or in a mass spectrometer in a lab on Earth. And they have already been successful in that. The first attempt was made by the Japanese space probe *Hayabusa*, which descended twice in November 2005 to the surface of the small NEO Itokawa and tried to sweep up some material there. On its return to Earth—after many technical setbacks and delays—it turned out it had only gathered about 1,500 microscopic grains, but it was a start.

The Japanese had more success with asteroid Ryugu; at 0.5 miles in diameter, this asteroid is twice the size of Itokawa. *Hayabusa-2* had a couple of mini landers and rovers on board that, in 2019, made a soft landing, hopped around on the rocky surface of Ryugu, and took a range of measurements. The space probe itself also descended, scooped up a few grams of surface material, and sent it back to Earth: the capsule with the Ryugu samples landed in Australia on December 5, 2020. *Hayabusa-2* then continued on through the solar system; it still has encounters with the small near-Earth objects 2001 CC_{21} and 1998 KY_{26} on its schedule.

The most successful sampling mission to date is NASA's *OSIRIS-REx*. The name is a convoluted acronym of Origins, Spectral Interpretation, Resource Identification, Security, and Regolith Explorer. Launched in summer 2016, it arrived in orbit around the 550-yard PHA (potentially hazardous asteroid) Bennu in late 2018. This asteroid has a 1 in 1,700 chance of colliding with Earth in 2182. An ingenious mechanism sucked up an estimated 4.3 ounces (just 122 grams) of asteroid material. This was then sealed in a capsule that landed in the Utah desert on September 24, 2023. *OSIRIS-REx* also conducted extensive geological and mineralogical research on the small celestial body, which (like Ryugu) is a spinning pile of rubble in the shape of two cones whose bases touch. The spacecraft (renamed *OSIRIS-APEX*) is now en route to an encounter with asteroid Apophis during its close encounter in April 2029.

The material we now have stored in laboratories on Earth has not only come from asteroids; dust from comets has also been collected for detailed study. On January 2, 2004, the US *Stardust* probe, launched in February 1999, flew through the coma of comet Wild 2—the tenuous cloud of gas and dust particles around the actual cometary nucleus. Using a kind of fly swatter, *Stardust* captured cometary dust, which was delivered to Earth in January 2006.

Space research on comets actually began back in March 1986, when no fewer than five spacecraft flew past the nucleus of the famous comet Halley: the Russian Venus explorers *Vega 1* and *2*, the Japanese probes *Sakigake* and *Suisei*, and the successful European *Giotto*. The latter took the first close-up pictures of a comet from some 370 miles away, while the space probe itself was sandblasted by the cloud of dust particles it flew through at high speed. *Giotto* was the first to image individual geysers on the surface of a comet—weak spots in the crust through which evaporated ice forces its way out, under the influence of solar heat.

Another spectacular comet mission was that of the American *Deep Impact* space probe, mentioned briefly in the previous chapter. On July 4, 2005, the probe fired an 815-pound copper projectile at the nucleus of comet Tempel 1, to study material from under its surface. The small crater produced by the impact was examined in 2011 by *Stardust*, which flew past it (years after its own Wild 2 adventure) at a distance of 125 miles. The *Deep Impact* project did not, by the way, lead to any noticeable change of Tempel 1's orbit. According to the researchers, it was like throwing a pebble at a passing 18-wheeler truck.[1]

By far the most impressive comet research program was that of the European *Rosetta* space probe, which was launched in March 2004. In 2014, after some convoluted looping through the inner solar system, *Rosetta* arrived at comet Churyumov–Gerasimenko, also known as the "rubber duck comet" due to its remarkable shape. For months, *Rosetta* flew along with the comet's approximately 3.5-mile-wide nucleus, studying how activity on its surface develops as it approaches the Sun. The small lander *Philae* descended to the surface on November 12, 2014, hopped around in the comet's extremely weak gravitational field, and finally came to a halt at the foot of an ice cliff where, after a few months, it finally gave up the ghost.

And while we're talking about studies of the minor inhabitants of the solar system, a mention of the American *New Horizons* mission is in order. *New Horizons* flew past the dwarf planet Pluto—an icy world with glaciers of frozen nitrogen—on July 14, 2015, after a journey of almost 9.5 years, and visited the ice dwarf Arrokoth, 4 billion miles from the Sun, on January 1, 2019. Like the *Rosetta* comet, Arrokoth was found to consist of two flattened lobes, in this case 13 and 9 miles in size, a shape that brings to mind a snowman.

And there is much more in store for us. The NASA probe *Lucy*, launched on October 16, 2021, will make a detailed study of no fewer than ten asteroids between 2027 and 2033 (two of which are orbiting each other), most of them Trojans (see chapter 2). During its first encounter, on November 1, 2023, *Lucy* discovered a tiny two-lobed moon orbiting asteroid Dinkinesh. And the 160-mile-wide asteroid Psyche, which has an iron-rich surface and may be the core of an ancient protoplanet, will be visited in August 2029 by the similarly named space probe that was launched on October 13, 2023. The United Arab Emirates also announced plans in the autumn of 2021 for research on seven asteroids and a landing on one of those seven.

This slim book is not the right place to discuss at length all the newest knowledge about the origin of the solar system that we already owe to asteroid research. Besides, astronomers themselves do not yet have a picture that is both complete and beyond doubt. The fact is that asteroids and comets are not only a potential threat to life on Earth but also a valuable source of information about the origins of Earth and the other planets. And that is not the only good news, as will be shown in the final chapter.

10
The Flip Side

If you can't beat them, join them. Sensible advice in many everyday situations, and, who knows, it may also apply to asteroids and comets that pose a potential threat to our planet. We have seen that it is impossible to totally avert the danger of cosmic collisions. Sooner or later Earth will be hit by an extraterrestrial projectile. But if we can't render the assailants completely harmless, we can at least try to get some benefit from them.

Certainly, having a meteorite land on you is no fun; being injured by flying glass after a serious atmospheric explosion is not something you would wish on anyone, and a really serious impact will undoubtedly cause death and destruction on Earth. But at the same time, we have to realize that, without such a cosmic disaster, we would not even be here, and that we even owe our oceans and the building blocks of life on Earth to collisions with other celestial bodies. As with the Hindu goddess Shiva, destruction and creation in the solar system often go hand in hand.

On a small scale, mankind has thousands of years of experience with converting something that is dangerous into a thing of beauty. Five-thousand-year-old "beads" made of meteorite iron have been discovered in Egypt. After the impact of a large meteorite in the Indian province of Punjab in 1621, Mughal emperor Jahangir had magnificent swords and daggers forged from the extraterrestrial material. In Indonesia, too, the finest kris (ceremonial daggers) are made of meteoritic iron; thanks to the high nickel content, the gracefully curving knives often have a silvery sheen. Bear in mind that terrestrial iron is almost always underground, in the form of iron ore, which means that in the distant past it was virtually inaccessible.

Early in the twentieth century, American geologist Daniel Barringer founded the Standard Iron Company, aiming to search the impact crater in Arizona named after him for the large iron deposits he expected to find there. Barringer had little success, but the iron and nickel mines in Sudbury, Canada, do owe their existence to a catastrophic cosmic impact in the geologically distant past. And large meteorites do not only contain a lot of iron and nickel (proportionally much more than Earth's crust because most heavy elements sank to the core shortly after Earth's formation); many rare metals, such as cobalt, palladium, platinum, and gold, are also present in larger quantities.

To give you an idea: a 25-yard-wide metal-rich asteroid can easily contain 10 tons of platinum, with a worth of hundreds of millions of dollars. And if you could trade all the iron, nickel, and precious metals in a large celestial body like Psyche (whose diameter is some 160 miles), it would fetch quintillions of dollars. It is no wonder, then, that there is a lot of interest in mining those colossal alien deposits. Mining in space could easily become a very profitable business in the future. Even extracting the large quantities of water in asteroids and comets could be very lucrative: hydrogen and oxygen are key components of rocket fuel, and if you can produce these elements in space by splitting up water molecules (a process known as electrolysis), it is far cheaper than transporting them from Earth with its strong gravitational field.

In 2012, scientists from the California Institute of Technology (Caltech) carried out a feasibility study on asteroid mining. A short while later, two commercial companies were set up in the US with the serious intention of working further on this: Planetary Resources and Deep Space Industries. The idea was at some point in the future to identify a suitable metal-rich near-Earth object that could be made to orbit around Earth using the deflection techniques described in previous chapters (a gravitational tractor, for example). A fully automated mining station would then slowly but surely excavate the asteroid, and the extracted minerals would be transported to Earth at regular intervals. This all sounds like science fiction, and it may come as no surprise that the ideas have so far never gone beyond the drawing board.

Planetary Resources was set up in 2012 in part by Peter Diamandis, who gained notoriety with his X Prize for innovation in aerospace technology. Four years later, Diamandis and his colleagues had amassed several tens of millions of dollars from wealthy venture capitalist investors, including Google cofounder Larry Page and movie director James Cameron. In 2015 and 2018 the company even gained some practical aerospace experience by launching two small test satellites. Unfortunately, the company is now defunct. Nothing has been heard from Deep Space Industries either. This is most probably an instance of the handicap of a head start: these true pioneers may not have succeeded in realizing their dream, but that does not mean that asteroid mining will never play an important role in the global economy. Or perhaps we should even say: in the solar system economy.

Whether or not we will be able to monetize potentially dangerous asteroids in the future, the fact is that we owe our existence to cosmic impacts. Clearly, we cannot repeat the experiment of biological evolution with different assumptions and starting conditions, but it seems highly unlikely that *Homo sapiens* would be walking the earth had it not been for the Chicxulub impact 66 million years ago. The extinction of the dinosaurs—and numerous other life-forms on land and in the oceans—marked a turning point in the history of life on Earth, and eventually provided the opportunity for the evolution of primates.

Even the water on our planet—indispensable for the origin of life according to most biologists—we owe to catastrophic collisions between Earth and other, smaller celestial bodies. In other words, almost all our water most certainly has an extraterrestrial origin. That may sound incredible, especially considering that over two-thirds of Earth's surface is made up of seas and oceans. However, on average, those oceans are no more than a few miles deep—a very thin layer compared to the diameter of Earth. Consequently, the mass of all Earth's water is only 0.023 percent of the total mass of the planet.

And the presence of water is not an obvious outcome. Shortly after Earth was formed—through the collision and coalescing of smaller protoplanets—it was in a sense one giant mass of glowing magma, and all the rock must have been completely boiled dry. It was only at a later stage, when the planet had cooled down enough, that an ocean of liquid water could exist on the surface. The questions are, of course, where that water came from, and especially when was it "delivered" to the newborn Earth.

For a long time, astronomers assumed that most of Earth's water had come from comets that had collided with our planet in large numbers during the solar system's youth. However, detailed chemical studies have since shown that this is not correct. Earth's ocean water contains a relatively small amount of deuterium (heavy hydrogen, with an atomic nucleus containing not only a proton but also a neutron), and with very few exceptions, the deuterium content of cometary ice is about twice that of Earth's water. So, that water must have come from another "reservoir"; cometary ice makes up at most 10 percent.

Nowadays, it is generally believed that almost all the water on our planet was derived from asteroids. These originated in another part of the solar system, at a shorter distance from the Sun, and they do have the same deuterium/hydrogen ratio as terrestrial ocean water. Research on asteroids has shown that they contain a lot of water (in the form of both ice and hydrated rocks), and we also know that there was a heavy interplanetary bombardment of small celestial bodies in the early infancy of the solar system—the Late Heavy Bombardment discussed previously.

If asteroids brought water to Earth long ago, they may also have delivered organic molecules—the building blocks of life. Chemists generally use the term "organic molecules" to refer to hydrocarbons and other compounds of hydrogen, oxygen, carbon, and nitrogen. These relatively complex molecules have been found in meteorites; the famous Murchison meteorite, which came down in the Australian state of Victoria on September 28, 1969, even contained dozens of different types of amino acids—essential building blocks of proteins. If it had not been bombarded by extraterrestrial projectiles, Earth might still be desolate and empty today.

At the same time, there must have been a rather delicate balance in the geological history of our planet.

Cosmic impacts may have made the emergence of life possible, but too many of these impacts could just as easily wipe that life off the face of the earth again. On that score, we can be thankful for the presence of Jupiter in the solar system. Computer simulations show that the impact frequency on Earth—particularly of comets—is strongly influenced by the gravitational influence of this giant planet. Without Jupiter, Chicxulub-like impacts on Earth would be at least 10 to 20 times more frequent.

Finally, the Moon also owes its existence to a catastrophic cosmic collision—an extraordinary fact when you realize that our comparatively large Moon has a stabilizing influence on the tilt of Earth's axis. It is generally believed that the Moon coalesced over 4 billion years ago from the debris of a collision between Earth and Theia—a hypothetical planet the size of Mars. Had that unimaginably violent collision not occurred, Earth wouldn't have the Moon it has today. Without the effect of the Moon, the tilt of Earth's axis would fluctuate enormously, as is the case with our neighboring planet Mars, and complex life might never have been able to develop.

*   *   *

On October 4, 2021, a person in Canada narrowly escaped serious injury from a meteorite impact. Hundreds of people saw a bright fireball in the sky that night, and Ruth Hamilton in Golden, a town in the province of British Columbia, was startled awake by a large, heavy space rock that drilled through the roof of her house and landed on the bed right next to her head. Hamilton escaped with just a scare; she was luckier than Ann Hodges, 67 years earlier in Alabama.

It will be clear by now that we are surrounded by large and small cosmic rocks. The day is bound to come when there will be further casualties from cosmic activity—either through someone being fatally hit by a hefty meteorite, or from the catastrophic consequences of a much heavier impact. We recognize the risks; we are alert and wary; and we prepare for such eventualities by developing potential countermeasures. But we are by no means safe. It is a humbling thought: despite our impressive scientific and technological capabilities, we are still fundamentally at the mercy of the whims of the cosmos—the same whims that made our existence possible.

Let's just be thankful that asteroids on a collision course, rapidly approaching comets, and other deadly projectiles are vastly rarer than the countless rocks, pebbles, grit particles, and flecks of dust that rain down on Earth every day. These are expertly captured and rendered harmless by our planet's atmosphere; every meteor and fireball we see appearing in the night sky bears beautiful and silent witness to this.

There is an old tradition that says if you see a falling star, you can make a wish. What would you wish for?

Notes

Chapter 1

1. Arthur Beiser, *The Earth* (New York: Time, 1962), 32.

2. O. Richard Norton, *Rocks from Space* (Missoula, MT: Mountain Press, 1994), 39.

3. Kevin Yau, Paul Weissman, and Donald Yeomans, "Meteorite Falls in China and Some Related Human Casualty Events," *Meteoritics* 29 (1994): 864.

Chapter 3

1. Ming Chen et al., "Yilan Crater, China: Evidence for an Origin by Meteorite Impact," *Meteoritics & Planetary Science* 56, no. 7 (July 2021): 1274–1292.

Chapter 4

1. Luis W. Alvarez et al., "Extraterrestrial Cause for the Cretaceous-Tertiary Extinction," *Science* 208, no. 4448 (June 6, 1980): 1095–1108.

2. J. Smit and J. Hertogen, "An Extraterrestrial Event at the Cretaceous–Tertiary Boundary," *Nature* 285 (1980): 198–200.

3. D. M. Raup and J. J. Sepkoski Jr., "Periodicity of Extinctions in the Geologic Past," *Proceedings of the National Academy of Sciences* 81, no. 3 (1984): 801.

Chapter 5

1. G. E. M. Gasper and B. K. Tanner, "'The Moon Quivered like a Snake': A Medieval Chronicler, Lunar Explosions, and a Puzzle for Modern Interpretation," *Endeavour* 44, no. 4 (2020).

2. W. F. Bottke and M. D. Norman, "The Late Heavy Bombardment," *Annual Reviews of Earth and Planetary Sciences* 45, no. 1 (2017): 619.

3. This catalogue has since been updated. For the latest version, see I. J. Daubar et al., "New Craters on Mars: An Updated Catalog," *Journal of Geophysical Research: Planets* 127, no. 7 (2022).

Chapter 9

1. Deep Impact, "FAQ," last updated November 14, 2017, https://deepimpact.astro.umd.edu/faq3.html.

Further Reading: More about Cosmic Projectiles

The Asteroid Threat: Defending Our Planet from Deadly Near-Earth Objects, William E. Burrows (New York: Prometheus, 2014)

Catching Stardust: Comets, Asteroids and the Birth of the Solar System, Natalie Starkey (London: Bloomsbury, 2018)

Collision Course! Cosmic Impacts and Life on Earth, Fred Bortz (Brookfield, CT: Millbrook, 2001)

Comet/Asteroid Impacts and Human Society, Peter Bobrowsky and Hans Rickman (eds.) (Berlin: Springer-Verlag, 2007)

Cosmic Impact: Understanding the Threat to Earth from Asteroids and Comets, Andrew May (London: Icon Books, 2019)

Rogue Asteroids and Doomsday Comets: The Search for the Million Megaton Menace that Threatens Life on Earth, Duncan Steel (New York: Wiley, 1995)

Index